Alternative materials in road construction

A guide to the use of recycled and secondary aggregates

P. T. Sherwood

T Thomas Telford

Published by Thomas Telford Publishing, Thomas Telford Ltd,
1 Heron Quay, London E14 4JD.
www.thomastelford.com

Distributors for Thomas Telford books are
USA: ASCE Press, 1801 Alexander Bell Drive, Reston, VA 20191-4400, USA
Japan: Maruzen Co. Ltd, Book Department, 3–10 Nihonbashi 2-chome,
Chuo-ku, Tokyo 103
Australia: DA Books and Journals, 648 Whitehorse Road, Mitcham 3132,
Victoria

First published 1995
Second edition 2001

A catalogue record for this book is available from the British Library

ISBN: 0 7277 3031 2

Typeset by Academic + Technical Typesetting, Bristol
Printed and bound in Great Britain by MPG Books, Bodmin

Contents

Acknowledgements

Particular thanks are due to Mrs. Sarah Groombridge of TRL for making the facilities of the TRL Library readily available and for patiently dealing with my enquiries. I would also like to thank Dr. R. J. Collins of BRE, Mr. A. Dawson of Nottingham University, Mr. B. Feldmann of BSI, Dr. J. M. Reid of TRL, Dr. L. K. Sear of UKQAA, Mr. D. York of Ballast Phoenix Ltd, and to thank COLAS Ltd for providing valuable information.

GENERAL INTRODUCTION

Since the first edition of this book was published in 1995 there has been increasing concern about the effect of man's activities on the environment which, if left unchecked, could lead to irreversible and possibly catastrophic changes to the life support systems on which we depend. The concern that such changes could occur has led to the concept of sustainable development which was defined in 1987 by the World Commission on Environment and Development as 'development that meets the needs of the present without compromising the ability of future generations to meet their own needs'. The UK government is committed to sustainable development and has identified four key aspects of development which need to be kept in balance if we are to achieve sustainability and a better quality of life. These are:

- maintaining a high and stable level of economic growth and employment
- social progress which recognizes the needs of everyone
- effective protection of the environment
- prudent use of natural resources.

Economic growth has inevitably led to increasing demands for aggregates for use in civil engineering construction. In the 20-year period to 1990 the total annual production within the UK of aggregates (sand, gravel and crushed rock) increased from 200 million tonnes to nearly 300 million tonnes. It was expected to continue to increase at a steady rate but has since fallen so that the 1997 level of 220 million tonnes was only 72% of the 1989 peak and production is now fairly stable at this level (Fig. 1).

Road building plays a significant role in this demand as it accounts for about one third of the total production (Fig. 2). On average 20 000 tonnes of aggregate are used for each mile of motorway construction. The environmental impacts of the extraction of aggregates are a source of significant concern across the country. These impacts include the loss of mature countryside, visual intrusion, heavy lorry traffic on unsuitable roads, noise, dust and blasting vibration. The extraction of aggregates also represents the loss of two finite natural resources – the aggregates themselves and the unspoilt countryside from which they are extracted.

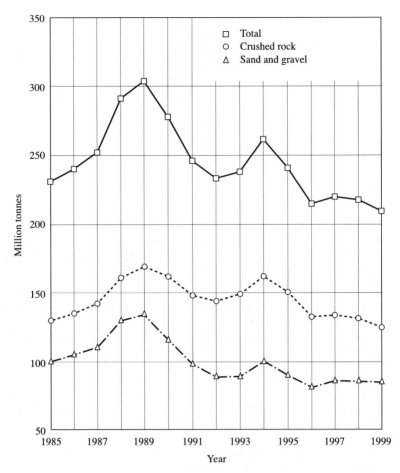

Fig. 1. Annual production of primary aggregates in the UK 1985–1999 (British Geological Survey 2000)

Concurrent with the production of aggregates, large amounts of waste materials and by-products are produced from industry and domestic use. The relative amounts and types of waste produced are shown in Fig. 3.

A summary of the production of mineral wastes made up of mining and quarrying wastes (25%) and construction and demolition wastes (16%), which form the subject of this book, is given in Table 1. Further details of the production and properties of these materials are given in later chapters.

If the materials listed in Table 1 could be used as alternatives to some of the materials used in civil engineering, their use would have the threefold benefit of conserving natural resources, disposing of the waste materials which are often the cause of unsightliness and dereliction, and clearing valuable land for other uses.

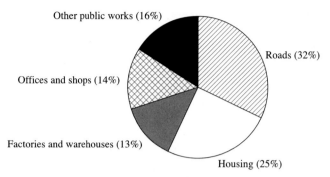

Fig. 2. Utilization of aggregates (BACMI 1991)

Although the use of alternative materials as secondary aggregates can have negative impacts on the environment (through transporting and reprocessing to produce secondary resources), their utilization has net environmental benefits because of the following.

- Recycling means that less waste is sent to landfill and other disposal options which have greater pollution and disamenity effects than recycling.
- The use of secondary aggregates reduces the environmental impacts of production through reduced energy use and reduced pollution.
- Using more secondary aggregates means using less primary aggregates which are non-renewable natural resources.
- The management, extraction, processing and distribution of primary aggregates can have environmental impacts that are greater than the environmental impacts of obtaining secondary aggregates from waste.
- Current rates of depletion of non-renewable natural resources may not be consistent with sustainable development.

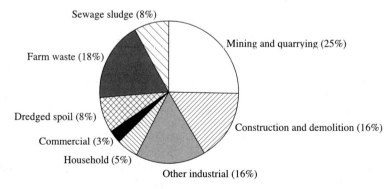

Fig. 3. UK Waste Production 1995. From 'Making Waste Work' (DOE Website 1995)

Table 1. Availability of major wastes and by-products with potential uses in road construction

Material	Annual production (million tonnes)	Year of estimate[a]	Stockpile (million tonnes)
Colliery spoil	13	1998	3600
China clay wastes	22	1998	300 (sand) 300 (other)
Power station wastes:	6.5	1997	250
Pulverized fuel ash	5.1		
Furnace bottom ash	1.4		
Blast furnace slag	4	1996	Some old tips
Steel slag	1.7	1996	12
Slate waste	9	1998	450
Spent oil shale	0	1991	150
Road planings	7	1991	0
Demolition wastes:	72.5	2000	0
Hard waste	33.8		
Soil	23.7		
Mixed	15.0		
MSWI[b] (incinerated refuse)	1.1	2000	0

[a]Most estimates derived from Digest of Environmental Statistics (DETR 2000)
[b]MSWI, Municipal Solid Waste Incinerated Ash

Diverting waste from landfill or tipping to recycling therefore substitutes a practice that has only environmental costs with a practice that has net environmental benefits. However, there are potential barriers to recycling which have been summarized (Reid 2000b) as:

- many older specifications do not permit the use of alternative materials
- many people are not aware of the methods available for recycling, developments in specifications and successful projects
- perception of use of alternative materials and recycling as being a high-risk activity
- where cheap natural aggregates are available, alternative materials may not be price competitive
- concerns about leaching of contaminants from alternative materials
- practical difficulties with individual materials and methods, often site specific.

This book therefore considers the extent to which waste materials and industrial by-products can be used in road construction in place of the natural materials that are traditionally used. Although it is difficult to avoid describing some materials as waste, the word has unfortunate connotations since it implies that the material has no use. Kwan and Jardine (1997) have pointed out that once a material is called 'waste' it enters a process from which it is difficult to re-invent it as a usable or valuable commodity. Once a construction use has been identified for a 'waste' material it is therefore preferable to refer to it as a recycled or secondary aggregate and this is the reason for the sub-title of this book.

The book is divided into three parts. Part I discusses the demands and requirements of roadmaking materials and the specifications that they must meet if they are to give satisfactory performance in each of the road pavement layers, from bulk fill at the bottom to the aggregates used in the surface layers at the top. Part 2 describes, in turn, each of the materials that are available as viable alternatives to the naturally-occurring materials that are traditionally used. It describes their physical and chemical properties and discusses their potential for use as road-making materials. Part 3 discusses the reasons why it is desirable that waste materials and by-products should be used in preference to naturally-occurring materials. It examines the optimum use of wastes and by-products, bearing in mind the need to minimize resource depletion, environmental degradation and energy consumption.

A Victorian view of the value of recycling and the penalties awaiting failure. The woman is saying 'You see, Richard, the wretched state to which you have brought your wife and children through your waste and improvident habits'.

© P.T. Sherwood 2001

PART 1
REQUIREMENTS FOR
ROADMAKING MATERIALS

If alternative materials are to be used in road construction they have to be classified and meet specification requirements in the same way that classification systems and specifications have been drawn up for roadmaking materials already in use. In theory, the specifications for alternative materials need not necessarily be the same as those for traditional materials, but in current practice this is usually the case, although it is rapidly changing. Before describing, in Part 2, the potential uses of the various alternative materials that are available, Part 1 therefore considers the requirements they can be expected to meet if they are to be seriously considered.

1. Classification and sources

Classification of roadmaking materials

Roadmaking materials can be classified either by the type of material, e.g. aggregate, cement, bitumen, etc. or by the pavement layer in which they are to be used, e.g. sub-base material. In the second example, which is more appropriate to alternative materials, it is necessary, before discussing the demand for, and specification of, roadmaking materials, to consider the methods currently in use for the vast majority of road construction projects.

Layer structure of roads

A road structure is made up of a number of layers (pavement layers) which are shown diagrammatically in Fig. 4. Two main types of construction are used:

(*a*) flexible, in which the top layers are bituminous-bound
(*b*) rigid, where the top layer is high-quality concrete.

There is also a third category known as composite construction, where the upper layers are constructed from bituminous material and are supported on a road base or lower road base of cement-bound material.

In a layered road pavement structure of the type considered here, the quality, in terms of durability and bearing capacity, of each of the pavement layers increases from the bottom upwards, i.e. the specification requirements for any given layer are always higher than those of the layer immediately beneath it. This means that the same material could be used for the construction of a particular layer and all the underlying layers; in principle, the whole road structure could be constructed from the materials used for the top layer. Full-depth asphalt construction goes some way to achieving this aim. However, building in layers generally means that costs are reduced and a very wide range of construction materials can be used. These materials range from material that occurs free on site for the construction of the bottom layer, to high-cost aggregates of high strength and skid resistance for the construction of the surface layer. This means that the scope for finding alternative materials as replacements for naturally-occurring materials decreases as the specification requirements for the respective layers increases. Bulk fill is therefore the biggest potential outlet for wastes, while very few such materials can be used in the surface layers.

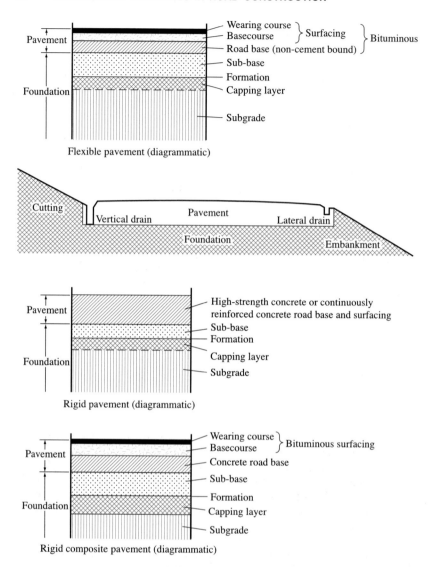

Flexible pavement (diagrammatic)

Rigid pavement (diagrammatic)

Rigid composite pavement (diagrammatic)

Unbound granular material can be used in all layers of the road foundation, including the sub-base and as a fill in the construction of cuttings and embankments. Surfacing and road base materials are generally bound in a less permeable matrix such as cement or bitumen.

Fig. 4. Principal features of a road construction (after Baldwin et al. 1997)

The bottom of the road structure on which the pavement layers are constructed is known as the subgrade. This may be in-situ material (usually soil), or fill material which has been imported to make up the level or to replace the in-situ material if this is too unstable to permit construction

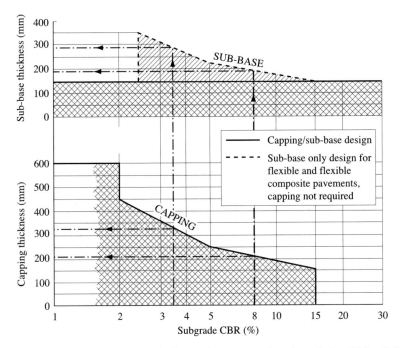

Fig. 5. Capping and sub-base thickness design as a function of the CBR of the subgrade (from MCDH Vol. 7, Section 2, Part 2 HD25/94 (1994))

to proceed. The subgrade plays an important role in the design of the road structure as its bearing capacity decides the thickness of the road structure above it. The bearing capacity of the subgrade is most frequently measured in terms of the California Bearing Ratio (CBR), which is an empirical test developed by the California State Highways Department for the evaluation of subgrade strengths. The procedure relates all materials back to a well-graded and non-cohesive crushed rock which is considered to have a CBR value of 100. An example of how the CBR value is used in practice is given in Fig. 5 which shows how the design thicknesses of the two layers immediately above the subgrade (i.e. the capping layer and the sub-base) are dependent on the CBR of the subgrade.

Bulk fill materials

If the required level of a new road is not the same as that of the ground over which it is built, the level has to be lowered or raised, processes known as 'cut' and 'fill' respectively. As far as possible the road engineer will try to balance the volumes of cut and fill, but inevitably there will be occasions when extra material has to be imported to the site. Moreover, the exercise of balancing 'cut' and 'fill' is complicated by the fact that it is sometimes difficult, at the design stage, to identify the soils which will be suitable for excavation and re-compaction on site. The suitability of some materials, such as clayey soils, depends on moisture content and the way they are 'worked'.

Imported fill may be required in very large quantities. The major requirements of imported fill are that it should be relatively easy to transport, to place and to compact. Once compacted all that is required of it is that it shall provide a stable bed, strong enough to receive the layer above it which may, depending on circumstances, be the capping layer or the sub-base. These conditions are fairly easy to meet and it is not difficult to find suitable materials; for this reason bulk fill provides by far the biggest potential outlet for the use of alternative materials.

Aggregates

Aggregates are required at all levels of the pavement structure except for the subgrade (the ground on which the road structure is founded), although their use for imported fill is not precluded. It is also possible, in favourable circumstances, to dispense with aggregates for the construction of the capping layer and sub-base by upgrading the subgrade by stabilization with lime or cement.

Large volumes of aggregate are consumed by road building programmes. This reached a peak in 1989 when 96 million tonnes were used (29% of total aggregate production) and the amount of aggregates used in road construction is likely to remain high, even if only to maintain the existing road network. Over and above the aggregates used in carriageway construction, large quantities are often required for ancillary works. In rural schemes the most important ancillary works, from the view point of aggregate usage, are bridges, drains, kerbs and verges. In urban schemes, footways, subways and other non-carriageway works become equally, or more, important. Please and Pike (1968) estimated that for major road construction 1.3 tonnes of aggregate per square metre of road pavement were required, while the ancillary works imposed an extra demand on average of about half of this amount. A more recent estimate (OECD 1997) is that 10 000 cubic metres of aggregate are needed for each kilometre of two-lane road construction.

The sub-base, capping and imported fill represent, in terms of volume (but not of cost), by far the greatest proportion of the road structure. Figure 5 shows that the combined thickness of capping and sub-base may be as much as 750 mm and is never less than 150 mm. Table 2 gives the estimated amounts of granular material used at capping layer and sub-base level for different thicknesses of construction.

To the aggregates that are used in the capping layer and sub-base must be added the materials used in the earthworks and related construction. These are not classified as aggregates and do not therefore appear in the statistics, so it is difficult to give a precise estimate of what this proportion might be. These materials may occur on site, but they frequently have to be imported in large quantities and can in themselves make up a significant proportion of the total amount of material used in road construction.

In contrast to the combined thickness of the lower layers, the greatest design thickness of the most heavily trafficked road would never exceed

Table 2. Estimated aggregate demand for various thicknesses of sub-base and capping of a three-lane motorway (Collins et al. 1993)

Thickness of capping layer[a] (mm)	Thickness of granular sub-base[a] (mm)	Thickness of capping and granular sub-base[a] (mm)	Weight of aggregate (tonnes/km)
0	150	150	13 000
0	225	225	20 000
350	150	500	43 000
600	150	750	65 000

[a] Thicknesses are those given in Fig. 5

a total thickness of 450 mm of road base and surfacing. Thus, although the specification requirements for bulk fill, capping and sub-base are less onerous than those for the upper layers, these foundation layers are likely to represent the bulk of the material requirements for road pavement construction. It is clear from this that the total thickness of the sub-base, capping layer and any imported fill is likely to be at least half of the total thickness of the whole road structure.

Binders
Soils, aggregates and related materials are used in an unbound condition in the lower layers of the road pavement. However, as they do not have sufficient stability for use in the upper layers they invariably have to be mixed with a binding agent which bonds the particles together by physical and/or chemical means. The binding agents of particular importance are bitumen, Portland cement and, to a much lesser extent, lime (calcium oxide CaO and calcium hydroxide $Ca(OH)_2$). The binder content is only a small proportion of a bound mixture so that, in comparison with the amounts of aggregates that are used in road construction (Figs 1 and 2), the amounts of binder used are relatively small, but still significant (Table 3).

Road construction accounts for most of the consumption of bitumen but less than one fifth of the consumption of Portland cement. The properties of both may be enhanced by the use of secondary additives but, in the case of bitumen, the scope for using any of the alternative materials considered in this book as an additive is almost non-existent. By contrast, considerable technical, economic and environmental benefits may be gained from the addition of pulverized fuel ash and blast furnace slag to Portland cement, and these are considered further in Parts 2 and 3.

Sources of roadmaking materials
Naturally-occurring materials
Naturally-occurring materials have traditionally been used for road construction and they still represent by far the biggest source of supply. The materials may occur on site, as in the case of the in-situ soil or,

Table 3. Production of Portland cement and bitumen in the UK 1988–1998 (Annual Abstract of Statistics 2000)

Year	Bitumen produced for inland consumption (thousand tonnes)	Cement production (thousand tonnes)
1988	2296	16 506
1989	2393	16 849
1990	2454	14 740
1991	2302	12 297
1992	2336	11 006
1993	2450	11 039
1994	2569	12 307
1995	2459	11 805
1996	2189	12 214
1997	2258	12 638
1998	2190	12 409

more typically, as in the case of sand, gravel and crushed rock, they may be quarried.

Until very recently most specifications for roadmaking materials were consequently based on the assumption that natural materials would be used. This is slowly changing as more emphasis is being given to recycling and to the use of alternative materials. But even on the most optimistic assumptions it is still presumed that naturally-occurring materials will make up the bulk. These materials may be used as they occur in nature, or they may be processed to produce a more consistent product, or they may be upgraded by the addition of a binding agent such as cement or bitumen.

Manufactured materials
Manufactured materials in this context refer to products that do not occur naturally but are specifically manufactured for use in building and civil engineering construction; they are distinct from those that are produced as by-products or wastes from other manufacturing process. Portland cement, bitumen and lime are the chief examples of manufactured materials used in significant quantities in road construction, and these have already been considered.

Apart from these, significant amounts of artificial aggregates are manufactured for use in civil engineering, for example to provide durable materials of light weight or with good insulating properties. Compared with natural aggregates, these materials are expensive and their use is justified only for those purposes where the benefits of using them outweigh the considerable increase in costs. For example, in high-rise buildings, concrete made from lightweight aggregate reduces the building weight with consequent savings in the size of the foundations. However, except possibly in the case of bridge construction and the production of materials with high skid resistance, there is little scope for using artificial aggregates in road construction.

Alternative materials (wastes, by-products and recycled aggregates)
As mentioned in the Introduction concern about the depletion of natural
resources and the effect that meeting the demand for aggregates may have
on the environment has increasingly focussed attention on the possibility
of finding alternatives to naturally-occurring materials. A number of
possible sources of supply have been examined, the most important
being the potential for use in road construction of the mineral waste
materials and industrial by-products that are listed in Table 1. Other
alternatives include the possibility of in-situ recycling of the road pavement layers and the manufacture of artificial aggregates with properties
similar to, or even superior to, natural aggregates.

Part 2 of this book discusses the potential uses of alternative materials
in road construction and in Part 3 the environmental benefits of using
them in preference to naturally-occurring materials are discussed. However, before this can be done the specifications and standards in use for
roadmaking materials need first to be examined, and this is the subject
of the next chapter.

2. Specifications and standards for roadmaking materials

British specifications

In the UK the requirements given in the Specification for Highway Works are mandatory for all road construction which is funded by central government. It was first published in 1951 and is revised at regular intervals; the current edition was published in 1998. It is a national specification which started life as a publication of what was then the Ministry of Transport. Several changes in the organization of government departments have since happened and the latest edition is a joint publication of the Highways Agency (for England), the Scottish Executive Development Department, the National Assembly for Wales and the Department for Regional Development for Northern Ireland. The Specification for Highway Works is Volume 1 of a multivolume publication – the Manual of Contract Documents for Highways (MCDH). These volumes are revised from time to time and do not all bear the same date.

A survey of the specifications of aggregates and bulk construction materials (Collins *et al.* 1993) showed that the Specification for Highway Works formed the basis of most other specifications in use in the UK for road and related construction. It also showed that where departures were made from the national specification this was more likely to lead to more stringent requirements being made rather than to any relaxations. The review concluded that the national specification allowed a wide range of materials to be used and that it could not be regarded as being unduly restrictive.

The evaluation of a material for any particular application in road construction is therefore most readily done by comparing its properties with those of materials known to be satisfactory, as described in the national specification. Given that the specification is so widely used as the basis for all road construction in the UK there is, at present, little alternative to this approach. However, it suffers from the fact that most of the specifications in use are recipe-type, i.e. the properties of a material to be used for a given purpose are closely defined and it is assumed that if a material meets all the specification requirements it will perform satisfactorily. The material properties that should be specified are based on years of experience that materials with such properties will perform satisfactorily. This method works well with traditional materials, but when new materials are introduced they may fail to meet the specification requirements even if they would in fact be perfectly satisfactory.

Research is in progress to define more closely the requirements of the individual pavement layers so that 'end-product' specifications can be used which would permit the use of any material that could meet the design requirements of a particular pavement layer. However, until the results of this research are incorporated into new specifications there is no solution other than to accept that if alternative materials are to be used it must be shown that they can meet the existing specifications. However, for minor roads judicious relaxation of the requirements given in the national specification is a possibility that could lead to greater use of marginal materials which do not fully comply.

Some of the more commonly occurring alternative materials are mentioned by name in the Specification and an Advice Note has been published by the Highways Agency which gives guidance on conservation techniques and the use of reclaimed materials and industrial by-products currently permitted. Table 4 is reproduced here from this Advice Note.

Table 4 gives a summary of the materials considered as being suitable or unsuitable for particular applications. Where this is the case the reasons for the inclusion or exclusion are discussed in the relevant part of this book. However, materials not mentioned may also be suitable or give rise to problems, and these cases are also considered.

Specifications for fill materials

The Specification for Highway Works 1998 (in the Design Manual for Highways and Bridges) gives a wide range of fill materials which are regarded as acceptable, and defines unacceptable materials as:

(*a*)   material from peat, swamps, marshes and bogs
(*b*)   logs, stumps and perishable material
(*c*)   materials in a frozen condition
(*d*)   clays having a liquid limit in excess of 90% or a plasticity index exceeding 65% (note: only a small minority of clays would fall into this category)
(*e*)   material susceptible to spontaneous combustion, except unburnt colliery spoil compacted in accordance with methods specified
(*f*)   material with hazardous chemical or physical properties.

A minor additional requirement is that when the fill is to be placed within 500 mm of concrete, cement-bound or other cementitious materials, the soluble sulphate content should not exceed 1.9 g/litre. Similarly, if the fill is to be placed within 500 mm of a metal structure, the total sulphate content should not exceed 0.5%.

Apart from common fill, the specification also includes requirements for special fills where aggregates are required. Recent editions of the Specification for Highway Works have shown an increasing emphasis on the use of alternative materials. The 1976 and earlier editions contained specifications for granular materials for use in earthworks that, for the most part, assumed that natural aggregates would be imported to the site. The 1986 and subsequent editions advise that the best use should be made of

Table 4. Permitted uses of secondary aggregates in the Specification for Highway Works (From Advice Note HD35/95)

Application	Embankment & Fill	Capping	Unbound Sub-base	Bitumen-bound layers	Cement-bound layers	Cement-bound road base	PQ concrete
SERIES	600	800	800	900	1000	1000	1000
MATERIAL							
Crushed concrete	A	B	A	C	A	A	A
Reclaimed bituminous materials	B	B	C	A	B	C	C
Demolition wastes	B	B	C	C	B	C	C
Blast-furnace slag	A	B	A	A	A	A	A
Steel slag	C	C	A	A	B	C	C
Burnt colliery spoil	A	B	A	C	B	C	C
Unburnt colliery spoil	B	C	C	C	B	C	C
Spent oil shale	B	B	A	C	B	C	C
PFA	B	A	C	C	B	A	A
FBA	B	B	C	C	B	C	C
China clay waste	B	B	B	B	B	B	B
Slate waste	B	B	B	C	B	B	B

Key
A: Specific provision
B: General provision – permitted if the material complies with the Specification requirements but not named within the Specification
C: Not permitted
Important note:
Materials indicated as complying with the Specification for a particular application may not necessarily comply with all the requirements of the series listed, only particular clauses. For example in the 600 series unburnt colliery spoil can satisfy the specification as a general fill but is excluded as a selected fill.

material arising on the site and that the aim should be to minimize the import of materials for economic and environmental reasons.

In the current specification these fills are listed as selected granular fill (Class 6) which is further sub-divided into further sub-categories 6A–6Q. The range of permitted constituents for these are summarized in Table 5. Limits for these in terms of grading, plasticity, organic matter content and sulphate content and particle strength are specified.

Specifications for capping layers
The functions of a capping layer are:

 (*a*)   to protect the subgrade from the adverse effects of wet weather

Table 5. Range of permitted constituents in earthworks materials (after Rockliff 1998)

Class	Sand & gravel	Crushed rock	Chalk	Argillaceous rock	Crushed concrete	Unburnt c. spoil	Burnt spoil	Slag	PFA
6A	✓	✓	✓	×	✓	×	✓	×	×
6B, C, D	✓	✓	✓	×	✓	×	✓	✓	×
6E	✓	✓	✓	×	✓	×	✓	✓	✓
6F1	✓	✓	✓	×	✓	×	✓	✓	✓
6F2	✓	✓	✓	×	✓	×	✓	✓	✓
6H, I, J	✓	✓	✓	×	✓	×	✓	✓	×
6K, L, M	✓	✓	✓	×	✓	×	✓	×	×
6N, P	✓	✓	✓	×	✓	×	✓	×	×
6Q	✓	✓	✓	×	✓	✓	✓	×	×

(b)   to provide a working platform on which the sub-base construction can proceed with minimum interruption in wet weather

(c)   to allow the full load-spreading capabilities of the sub-base to be realized, which would not be possible were it to be laid directly on top of a weak subgrade.

In recent years, the Department of Transport has encouraged the stabilization of the in-situ subgrade material for the construction of capping layers (Wood 1988). The alternative to stabilization of the in-situ subgrade is either to use unbound granular material or to stabilize with cement imported granular material which would be unsuitable for use in an unbound form. The specification recognizes the following granular materials for capping.

Class 6E. Selected granular material for stabilization with cement, which may be any material, or combination of materials, other than unburnt colliery spoil and argillaceous rock.

Class 6F1. Selected granular material (fine grading) which may be any material, or combination of materials, other than unburnt colliery spoil.

Class 6F2. Selected granular material (coarse grading) which may be the same types of material as are specified for Class 6F1.

The requirements for these materials are summarized in Fig. 6. Although it is not written into the specification, it can be inferred from other publications that the materials when placed should have a minimum California Bearing Ratio (CBR) value of 15%. The CBR value is not an onerous requirement for either the unbound or cement-stabilized granular capping, and it can be seen from Fig. 6 that the grading requirements for unbound capping are very undemanding. Even if the grading requirements cannot be met a material may still be used if it satisfies the more relaxed grading requirement for cement-stabilized granular capping.

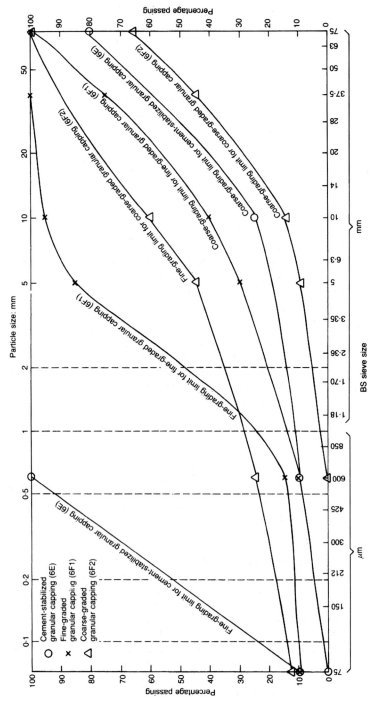

Fig. 6. Requirements for granular materials used in capping layer construction (DMRB. Specification for Highway Works 1998)

Other requirements	Class 6E	Class 6F1	Class 6F2
Liquid Limit (%)	<45	NR	NR
Plasticity Index (%)	<20	NR	NR
Organic content (%)	<2	NR	NR
Total sulphate (%)	<1	NR	NR
10% fines value (TFV)	NR	>30	*

NR: No requirement
* Value written into contract document

Fig. 6. Continued

Specifications for sub-bases
The functions of a sub-base are:

(a) to provide a working platform on which the paving materials can be transported and compacted

(b) to be a structural layer which assists in spreading the wheel loads so that the subgrade is not overstressed

(c) to be an insulating layer against freezing where the subgrade is a material likely to be weakened by frost action (to fulfil this function the sub-base must itself be frost-resistant).

The thickness of the sub-base is not related to traffic intensity, and only to a slight extent to the bearing capacity of the subgrade; deficiencies in this are compensated for by requiring a greater thickness of capping layer. At sub-base level both unbound and cement-bound aggregates may be used. The specifications for these are discussed below.

Unbound sub-base and base materials. Unbound aggregates are described in Clauses 803 and 804 of the Specification which give, respectively, the requirements for granular sub-base material Type 1 and granular sub-base material Type 2. Until recently there was an additional clause for wet-mix macadam but this is little used and it has now been excluded from the Specification. Type 1 granular materials include 'crushed rock, crushed slag, crushed concrete or well-burnt non-plastic shale'. Type 2 granular materials include the materials permitted for Type 1 and also natural sands and gravels. Wet-mix macadam was restricted to crushed rock or crushed slag. The major requirements of Type 1 and Type 2 granular sub-base material are summarized in Fig. 7.

Type 1 materials, which exclude the use of sands and gravels, are expected to be all-weather sub-base aggregates; Type 2 materials have a lower specification and are not expected to give good performance under construction in wet weather. Unbound sub-bases are not permitted in rigid pavement construction; they may be used in flexible pavement construction, but Type 2 materials are excluded if the design traffic loadings are more than 400 commercial vehicles per day.

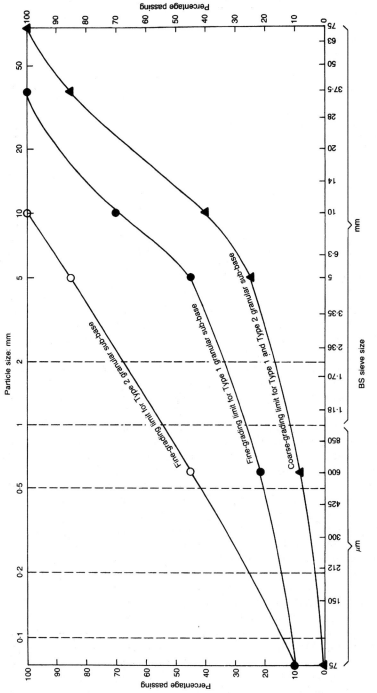

Fig. 7. Requirements for unbound granular materials used in sub-base construction (DMRB. Specification for Highway Works 1998)

Other requirements	Type 1 sub-base	Type 2 sub-base
Plasticity Index	0	<6
Soaked TFV (kN)	50	50
Soundness value	>75	>75
Water absorption (%)	<2	<2
Minimum CBR	*	30

* CBR assumed to be adequate
In addition to the other requirements unbound materials used within 450 mm of the road surface also have to be non-frost-susceptible when tested by the BS frost-heave test (BS 812 1988).

Fig. 7. Continued

Cement-bound sub-bases.   Cement-bound materials may be used as an alternative to unbound granular materials under flexible construction and are the only permitted materials under rigid construction. The Specification contains clauses for a family of cement-bound materials referred to by their initials as CBM1, CBM2, CBM3 and CBM4 (these were previously known, respectively, as soil–cement, cement-bound granular material and lean concrete). All could be used at sub-base level but the superior quality of CBM3 and CBM4 means that they are used principally for base construction or for sub-base under rigid construction while CBM1 and CBM2, described respectively in Clauses 1036 and 1037, are confined to sub-base construction.

The materials permitted for use for CBM1 and CBM2 are not specified as such, it being assumed that as long as they satisfy all the requirements of the specification they may be regarded as suitable. This means that, potentially, a very wide range of materials may be used. The materials to be used for the superior quality CBM3 and CBM4 are specified as natural aggregates complying with the requirements of BS 882 (1992) or air-cooled blast-furnace slag complying with the requirements of BS 1047 (1983), which complies with the quality and grading requirements of BS 882 (1992). The requirements of the CBM group of cement-bound sub-base and base materials which have to be satisfied are summarized in Fig. 8.

As an alternative to the CBM group of materials, which are compacted by rollers, the specification also includes a range of four cement-bound sub-base and base materials known as 'wet lean concrete 1–4' which are compacted by vibrator methods.

Cement-bound sub-bases under rigid construction.   Unbound sub-bases are not permitted to be used under rigid (concrete) construction, and the weaker cement-bound and wet lean concretes are also excluded. There are good reasons for this as the concrete paving is laid directly on to the sub-base (see Fig. 5) so that in effect the distinction between sub-base and base, which applies to flexible construction, does not exist. The sub-base/base has to provide a rigid platform on which the high-quality concrete paving can be laid. The requirements cannot therefore be regarded as

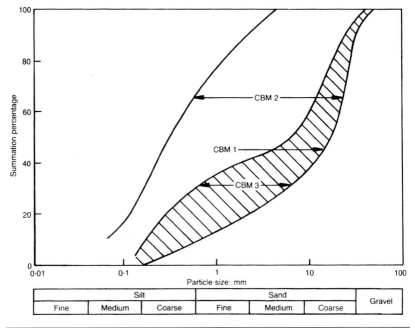

Other requirements	CBM1	CBM2	CBM3	CBM4
Soaked TFV (kN)	NR	50	NR	NR
Average 7-day compressive strength (MPa)	4.5	7.0	10.0	15.0
Minimum 7-day compressive strength (MPa)	2.5	4.5	6.5	10.0
Strength of 7-day air-cured and 7-day soaked as a percentage of 14-day air-cured	80	80	80	80

Fig. 8. Requirements for cement-bound sub-base and base materials. (The grading requirements for CBM1 relate to the coarse side of the grading envelope; there is no fine limit. The requirements for CBM3 and CBM4 are the same.) (DMRB. Specification For Highway Works 1998)

restrictive as it would be a false economy to risk the failure of expensive concrete paving by taking risks with the quality of the material on which it was laid.

Specifications for base construction
The road base must not be confused with the base course, which is an integral part of the surface course. The base course is a sub-layer within the bituminous surfacing; the road base is normally the thickest element of the flexible pavement on which the surfacing rests.

From a structural aspect, the road base is the most important layer of a flexible pavement. It is expected to bear the burden of distributing the applied surface loads so that the bearing capacity of the subgrade is not exceeded. Since it provides the pavement with added stiffness and resistance to fatigue, as well as contributing to overall thickness, the material used in a road base must always be of reasonably high quality. For this reason the scope for using alternatives to the naturally-occurring materials is severely restricted. The requirements of road base materials are not therefore considered here but are considered in Part 2, which deals with specific alternatives when these alternatives have some potential for use in road base construction.

Specifications for surfacing
The surfacing, and particularly the wearing course, represents only a small proportion of the total depth of construction. However, it accounts for a high proportion of the total cost of the pavement. Specifications for surfacing therefore impose more stringent requirements than is the case for the lower layers of the road pavement. The potential for using alternative materials in the surfacing is therefore very restricted and is considered, as appropriate, in Part 2.

Other national specifications
A survey of specifications in use for aggregates (Collins *et al.* 1993) showed that the specification of roadmaking aggregates in other countries did not differ significantly in fundamental principles from those used in this country. All countries adopt a layer technique (Figs 4 and 5) for road construction and use similar (but not identical) tests to determine suitability. This section reviews some of the national specifications that are in use.

European specifications
A collaborative research project with the acronym ALT-MAT was funded by the European Commission under the Transport RTD Programme of the 4th framework Programme in the mid-1990s. The work of this project was co-ordinated by the UK but most Western European countries were represented on the co-ordinating committee. The terms of reference of the project were to promote:

- improvements to the performance of all highway materials
- their efficient use and re-use
- the development and use of new materials
- the wider use of alternative materials in road construction
- a reduction in the consumption of scarce natural aggregates
- a reduction in the environmental impact of the disposal of the alternative materials.

The aim of the project was to provide information to bridge the gap between laboratory tests and field behaviour – to define methods by which the suitability of alternative materials for use in road construction could be evaluated under appropriate climatic conditions. The methods covered the mechanical properties, functional requirements, leaching potential and long-term stability of the materials and concentrated on unbound granular materials.

Much of the work undertaken by this project is more appropriately considered in Part 3 but it included an examination of the technical

Table 6. Overview of the technical requirements, national standards and/or additional requirements for alternative materials (ALT-MAT 1999)

Country	National requirements and standards	Materials with additional/ other requirements	Notes
Austria	×	• Steel slag • Blast-furnace slag	• No difference between natural and recycled materials concerning test methods or requirements • The quality of recycled building residues is regulated in guidelines
Denmark		• Crushed asphalt, as unbound road base • Crushed bricks, as unbound road base • MSWI bottom ash, as unbound sub-base	• For materials other than crushed asphalt, crushed bricks and MSWI bottom ash, Danish Road Standards are missing
Finland	×		• Studies on secondary materials of potential use is going on
France	×	• Blast-furnace slags	• Specific classifications of materials for different applications • If environmental preconditions exist, any material has to fulfil these
Sweden	×[a]		• No specific requirements on alternative materials. Residues may be used if they have at least as good properties as the material they replace
Switzerland	×		• Eight recycling products for road construction, for which applications are suggested and suitability tests listed (except for crushed glass and crushed concrete, all materials originate from roads)
United Kingdom	×		• Materials permitted for different applications are listed, also materials not permitted are listed[b] • Materials not listed may be permitted if they conform to appropriate standards
United States	× – Requirements differ among the states		
Japan	×	• Steel slag	

[a] For all roads constructed by the National Road Administration
[b] Primarily valid for the motorway and trunk road network, the responsibility of the Highways Agency

requirements and specifications used for roadmaking materials among the countries taking part in the project. Technical specifications for road construction in most countries used the same tests for natural and alternative materials. The alternative materials were assessed on the basis of the natural materials they most closely resembled. In some countries, a distinction was made between road by-products, which could be recycled for the same application with minimal testing, and non-road by-products, for which a comprehensive testing programme was required. Table 6 shows an overview of the technical requirements.

Specifications in the USA
  Unbound bases and sub-bases.   Although each state in the USA is a completely independent highway authority, the American Association of State Highway and Transportation Officials (AASHTO) and the American Society for Testing Materials (ASTM) have both issued specifications for unbound sub-base and base materials. These are revised at regular intervals and the last two digits of the specification represent the year in which it was revised. Thus ASTM Specification D2940-98 means that it was last revised in 1998.

AASHTO specification M147-65 'Materials for aggregates and soil–aggregate sub-base, base and surface courses' includes six types of aggregate, designated A–F. These differ with respect to their gradings (Table 7). Grading A is used primarily for bases and gradings B–D refer to sub-base materials. Gradings E and F are used as top courses for unsurfaced roads and have no counterparts in the UK. The requirements for plasticity and strength, as measured by the Los Angeles Test, are also given in Table 7. Material A has a grading very similar to that of the British Type 1 sub-base material, and the extremes of the grading envelopes of materials B, C and D correspond quite closely to Type 2 sub-base material (see Fig. 7).

Table 7. AASHTO requirements for unbound sub-bases and base materials (AASHTO M147-65)

Sieve size	Grading percentage passing					
	A	B	C	D	E	F
50 mm	100	100	100	100	100	100
25 mm	NR	75–95	100	100	100	100
9.5 mm	30–60	40–75	50–85	60–100	100	100
4.75 mm	25–55	30–60	35–65	50–85	55–100	70–100
2 mm	15–40	20–45	25–50	40–70	40–100	55–100
425 μm	8–20	15–30	15–30	25–45	20–50	30–70
75 μm	2–8	5–20	5–15	5–20	6–20	8–25

Other requirements
Liquid Limit of <425 μm fraction not greater than 25%
Plasticity Index of <425 μm fraction not greater than 6%
Percentage wear by Los Angeles Test not greater than 50%

Table 8. ASTM requirements for unbound sub-bases and base materials (ASTM D2940-98)

Sieve size	Graded percentage passing	
	Bases	Sub-bases
50 mm	100	100
37.5 mm	95–100	90–100
19 mm	70–92	NR
9.5 mm	50–70	NR
4.75 mm	35–55	30–60
600 μm	12–55	NR
75 μm	0–8	0–12

Other requirements
Coarse aggregate to be clean, hard and durable
Fraction passing the 75 μm sieve not to exceed 60% of fraction passing the 600 μm sieve
Liquid Limit of <425 μm fraction not greater than 25%
Plasticity Index of <425 μm fraction not greater than 4%

Two gradings, one for sub-base and the other for base, are specified by ASTM D2940-98 'Specification for graded aggregate material for bases and sub-bases for Highways and Airports' and, with other requirements specified by the ASTM, are given in Table 8.

Comparison of the ASTM requirements with those of their British counterparts, given in Fig. 7, shows that the sub-base grading is more restrictive, particularly with regard to the finer grading limit, than Type 2 sub-base, but the grading for base material is very similar to the British Type 1 granular sub-base.

Many states incorporate some of the AASHTO and ASTM requirements into their specifications. However, with the large variations in geology and climate, the individual specifications used by the independent state highway authorities differ significantly among the states. Geological differences play an important role: some states, such as Georgia and North Carolina, have abundant supplies of crushed rock and therefore prohibit the use of gravels. Others, with extensive gravel deposits, do not restrict the use of gravel. In general, however, where gravel is permitted most states have a requirement that all gravel particles must have at least one face fractured by crushing. All the states give a grading requirement and also have requirements for durability and the amount of plastic fines. The Los Angeles Test is most widely used for specifying durability; restrictions on the amount of plastic fines are imposed by specifying a maximum Plasticity Index or sand equivalent value.

Cement-bound bases and sub-bases. As in the case of unbound materials, each state has its own specification for cement-stabilized material used in the different layers of the road pavement but, in so far as there is a national specification, the comprehensive recommendations published by the Portland Cement Association (PCA) (1971, 1977 and 1979) apply. The philosophy adopted by the PCA is that a hardened stabilized material should be able to withstand exposure to the elements. The types of material to be used and their gradings are expressed in very broad terms; strength is considered to be a secondary requirement as most stabilized mixtures that possess adequate resistance to the elements also possess adequate strength.

The tests used to determine performance are the ASTM wetting and drying test (ASTM D559-96) and the ASTM freezing and thawing test (ASTM D560-96). These are designed to simulate what happens in practice, and provided a stabilized material can meet the requirements of both tests it is considered to be suitable for use regardless of its origins. In this respect the philosophy is not unlike that of the British cement-bound material category I (CBM1) where few requirements for the material to be used are given.

Standard methods of test for roadmaking materials

All the specifications for construction of the road pavement layers, considered earlier, require that the materials should be suitable for the purpose in mind. This means that they have to be tested before use and standard methods of test have therefore been developed which in the UK are published by the British Standards Institution (BSI). Counterparts to the BSI in most other countries also publish their own national standards. A survey of specifications used for aggregates (Collins *et al.* 1993) showed that the specifications of roadmaking aggregates did not differ significantly in fundamental principles from those used in this country. All countries adopt a layer technique (see Figs 4 and 5) for road construction and use similar (but not identical) tests to determine suitability.

An unpublished OECD survey in 1981 showed that with regard to unbound aggregates there was usually a grading, a strength requirement and a requirement to restrict the amount of plastic fines. The tests used to define these properties differed from country to country and even when they had the same name there were often subtle differences in the procedures, which meant that they did not have comparable results.

British Standards
With the exceptions mentioned later, there are no British Standards for alternative materials and by default they have to be tested by methods developed for aggregates and soils. The important standards in this respect are BS 812 'Testing aggregates', BS 1377 'Methods of tests for soils for civil engineering purposes' and BS 1924 'Methods of test for stabilized soils'. Each of these exists in more than one Part with BS 812 having no fewer than 24 Parts published over the period 1985 to 1989, each describing the procedure for a particular test.

Because of their high potential for use in construction there are several British Standards for blast-furnace slag and pulverized fuel ash. BS 1047 'Air-cooled blast-furnace slag for use in construction' was first published in 1942 and has since been revised on several occasions; the current edition was published in 1983. BS 3892: Part 1 'Specification of pulverized fuel ash for use with Portland cement' was published in 1997. Part 2 of this standard 'Specification of pulverized fuel ash to be used as a Type 1 addition' was published in 1996 and Part 3 'Specification of pulverized fuel ash for use in cementitious grouts' was published in 1997.

Apart from blast-furnace slag and pulverized fuel ash (PFA) there are no specific British Standards relating to alternative materials. However, BS 6543 'Guide to the use of industrial by-products and waste materials in building and civil engineering', published in 1985, gives guidance on the test methods which are most applicable.

At the time of writing (2001) all the British Standards mentioned here and in the References are still in force. However, it is expected that all National Standards in the European Union will gradually be withdrawn after 1 December 2003 and replaced by the European Standards discussed in the next section.

European Standards
In preparation for the single European market, the Comité Européen de Normalisation (CEN or European Committee for Standardization) has been preparing European Standards which will cover the use of all materials used in road construction, including the use of waste materials and by-products. CEN's ultimate objective is to produce fully harmonized European Standards, denoted by EN, which will replace existing national standards, such as those produced by the British Standards Institution.

Several standard methods of test have been or are being prepared to form a package of standards for aggregates but none will supersede any existing British Standards until December 2003. However, some have already been published by the BSI and are running concurrently with the corresponding British Standards for the time being. These have the designation 'BS EN' and when complete will describe many more test procedures than are at present covered by British Standards. In addition to the ENs, which correspond to BS specifications, there is a set of pre-standards designated as 'prEN' which correspond to BS Drafts for Development. If a pre-standard gains an overall majority vote, it is not necessary for conflicting national standards to be withdrawn and they may operate in parallel with the pre-standard. Like BS Drafts for Development, pre-standards have a finite life and are subject to reconsideration after a period of three years, when they can either be upgraded to full EN status or continue, in existing or modified form, for two years before further consideration.

The number of European Standards for aggregates and related materials will greatly exceed the number of British Standards and some will give specific requirements for end use rather than, as in the case of most British Standards, describing the test methods to be used. Requirements for end uses of aggregates will be specified in the following European Standards:

- prEN 12620 Aggregates for concrete

Table 9. Progress (2001) on the preparation of European Standards on Construction Products

Year	pre '93	'94	'95	'96	'97	'98	'99	'00	'01	'02	'03	'04	'05	'06
Under development	320	275	222	241	237	275	303	224	136	66	26	1	0	0
Under approval	106	125	157	205	209	290	292	307	260	238	179	131	66	26
Ratified	21	59	92	156	204	328	426	504	639	731	830	903	969	1009
Mandated	447	459	471	602	649	889	1017	1035	1035	1035	1035	1035	1035	1035

- prEN 13043 Aggregates for bituminous mixtures and surface treatments for roads and other trafficked areas
- prEN 13139 Aggregates for mortar
- prEN 13242 Aggregates for unbound and hydraulically bound materials for use in civil engineering work and road construction
- prEN 13383 Armourstone
- prEN 13450 Aggregates for railway track ballast.

The total number of BS EN Standards which have been or are being prepared are too numerous to enumerate here and in any case the position is constantly changing. However, where relevant, reference is made to them in the various sections of Part 3. Nor is there, at present, a formal catalogue of European Standards available. The nearest thing in the UK is the BSI catalogue of standards which, apart from listing all current British Standards, also lists BS ENs and prENs that have already been published. Table 9 gives a breakdown of the position of the preparation of European Standards relating to *all* construction products as of March 2001.

Approval of draft European Standards is carried out by a formal vote among the member countries. If the necessary overall majority is obtained, after counting weighted votes, all countries are obliged to adopt the EN to replace their national standards regardless of their vote. As in the case of the British Standards Institution, CEN operates through a series of Technical Committees (TCs) and Sub-Committees (SCs) which in turn may appoint Working Groups (WGs) and Task Groups (TGs). Each member country is represented on the Technical Committees and may, if it wishes, be represented on the Sub-Committees, but membership of Working Groups and Task Groups is usually confined to individual experts in the particular area of interest.

PART 2
ALTERNATIVE MATERIALS AVAILABLE – QUANTITIES, LOCATIONS, GENERAL PROPERTIES AND POTENTIAL USES

This part considers the alternative materials that are available in the UK and have some potential for use in road construction and maintenance. Each of the major materials listed below is considered in a separate chapter of this part. The listing is solely on an alphabetical basis and has no bearing on the relative suitability of the materials under discussion.

Each chapter describes the sources, the quantities currently produced and the quantities available for past production of the particular material. The chapters go on to consider the physical and chemical properties of the relevant material and end with a discussion of its potential uses in road construction and maintenance.

- Chapter 3. China clay wastes
- Chapter 4. Colliery spoil
- Chapter 5. Construction and demolition wastes

  ○ Recycled road pavement materials

- Chapter 6. Glass waste
- Chapter 7. Municipal waste
- Chapter 8. Power station wastes
- Chapter 9. Rubber
- Chapter 10. Slags

  ○ Blast-furnace slags
  ○ Steel slag
  ○ Non-ferrous slags
  ○ Imported slags

- Chapter 11. Slate waste
- Chapter 12. Spent oil shale

3. China clay wastes

Occurrence

China clay by-products have properties similar to those of primary aggregates and as such are an excellent source of roadmaking material. They are available in large quantities in south-west England but high road transport costs coupled with limitations in existing rail links mean that their transport to other parts of the country is uneconomic under present circumstances (AAS 1999).

China clay (kaolin) is used in the paper and ceramic industries. Most of the world's production is in south-west England and in North Carolina and Georgia in the USA, Britain being by far the largest producer. In south-west England, the commercial extraction of china clay (kaolin) from kaolinized granite is concentrated in the St Austell area with subsidiary workings in the nearby Bodmin and Lee Moor areas.

The kaolin was formed geologically by the action of steam and carbon dioxide on the orthoclase feldspar as the granite cooled and is extracted from steep-sided open pits by subjecting the face to high-pressure jets of water (see Fig. 9). The broken-up rock flows in a slurry to the pit bottom from where it is pumped to a separating plant. Here the bigger grains, which are predominantly quartz with small but variable amounts of other minerals, including a few flakes of mica, are separated from the sand waste. The residual slurry is dewatered and a second separating process removes the fine clayey sand and mica residue. Thus, for each tonne of china clay produced, about 9 tonnes of waste are also produced. This waste material is composed of approximately:

- 2 tonnes of overburden
- 2 tonnes of waste rock (stent)
- 3–7 tonnes of coarse sand waste
- 0–7 tonnes of micaceous residue.

Current mining regulations and other factors tend to make the extraction of china clay at depths much in excess of 45–60 m uneconomic, although the kaolinization continues to greater depths. Deeper workings are technically feasible and to avoid sterilizing these reserves from possible future reworking, the pits are not normally back-filled. Further, the production of kaolins to special requirements often needs the blending of kaolins from different sources and this is facilitated if the pits remain available for sporadic reworking as required.

Fig. 9. China clay workings

Where suitable, the pits are used for storing water, which is required in large quantities for extracting the china clay. The wastes are tipped, usually on land less suitable for china clay working. In the past all the waste residues were deposited on the same tips, which is unfortunate because the coarse sand waste has much more potential for use than do the other materials. More recently some of the granular material has been tipped separately, making it easier to extract.

Gutt *et al.* (1974) estimated that the total stockpile of wastes on the tips was 280 million tonnes. A more recent estimate gives a figure of 350 million tonnes for the stockpile (Whitbread *et al.* 1991). The stockpile continues to grow as the current production of wastes is estimated (Whitbread *et al.* 1991) to be 27 million tonnes per year and very little is used. The china clay industry is thus second only to coal mining as an originator of industrial wastes in the UK and it has the third largest stockpile; only the stockpiles of colliery spoil (3600 million tonnes) and of the almost defunct slate industry (400 million tonnes) are comparable in quantity.

Additionally, it is the most concentrated stockpiling of the industrial waste materials and a report commissioned by the Department of the Environment (1991b) concluded that: 'The china clay industry has particular problems with its very high level of permanent waste which it cannot return to the excavation. The tips are unsightly to many people ... and represent major issues which have yet to be resolved'.

Table 10. Chemical composition of china clay sand (Hocking 1994, OECD 1977)

Component	Composition (%)
SiO_2	75–90
Al_2O_3	5–15
Fe_2O_3	0.5–1.2
TiO_2	0.05–1.2
CaO	0.05–0.5
K_2O	1–7.5
Na_2O	0.02–0.75
MgO	0.05–0.5
Loss on ignition	1.2

Although 30–40% of Cornwall County Council's aggregate requirements are met by using china clay wastes, the demand for the wastes is insignificant compared to the annual output. As china clay production continues the stockpile is expected to increase at a substantial rate (Hocking 1994).

Composition of china clay wastes
Sand
Of the wastes produced by the extraction of china clay, the coarse sand is not only the largest component, but also that with the most desirable engineering properties. The sand is largely composed of quartz (85–88%), with smaller proportions of tourmaline (9–11%), feldspar (2–3%) and mica (1–2.5%). It is a quartzitic sand which is chemically inert; even if it did contain any soluble salts, most would be removed during the extraction process and any that remained would be rapidly leached out of such a free-draining material. A typical chemical analysis is given in Table 10.

The particle size of the sand is given in Fig. 10. The gradings of sand vary from tip to tip but it is possible to choose sands that meet the specifications for many roadmaking purposes.

Stent
The stent, which can vary in size from <100 mm to in excess of 2 m in diameter, essentially consists of massive quartz, quartz/tourmaline and partially kaolinized granite. The irregular distribution of these materials within the rock mass inevitably gives rise to variability in the waste. Typical properties reported by Hocking (1994) are:

- saturated TFV 75–150 kN
- magnesium sulphate soundness value 70–90%
- soluble sulphate content <0.1%.

Uses of china clay sand in road construction
Due to its abundance and superior properties, china clay sand is the most important of the waste materials of china clay production. The sand is a

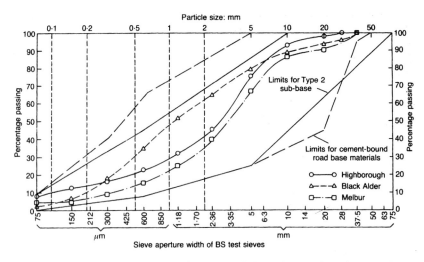

Fig. 10. Particle size distribution of samples of china clay sand (Tubey 1978)

good-quality aggregate which needs only the same basic grading and washing processes that are applied to other natural aggregates. It has accumulated in stockpiles only because of the large amounts produced in an area where there is low demand.

Bulk fill
At first sight china clay sand would appear to be an excellent material for bulk fill, but there has been a reluctance to use it in road construction. The reason for this is that micaceous materials have a dubious reputation among road engineers and the literature contains reports of compaction difficulties associated with their use. The obvious presence of mica in china clay sand has, therefore, given rise to suspicions that similar difficulties may occur when they are used in road works. However, compaction tests on selected sands (Tubey 1978) suggested that the mica present did not have any serious deleterious effects on the compaction.

Changes in density can be caused by the presence of mica. The larger change is produced solely as a result of the change in grading produced by the presence of mica. The other change is due to some physical property which is unique to mica; this is almost certainly the resilience of the mica flakes which allows them to deform under load and to recover after the load has been removed, rather like the leaves of a leaf spring. Research by Tubey and Webster (1978) on the effect of mica on the compaction properties of a range of materials showed that the resilience of the mica reduced the state of compaction achievable for a given compactive effort by about $0.007 \, \mathrm{Mg/m^3}$ for each one per cent of fine and coarse mica, respectively.

The distinctive colour, lustre and thin flaky shape of mica makes its presence, even in trace amounts, very obvious. It is possible, therefore, that many of the difficulties attributed to its presence in the past may have been caused by other factors, such as overall particle size distribution.

Despite the alleged problems, china clay sand has been used with considerable success as bulk fill for earthworks where a strict moisture content control was imposed to enable adequate compaction to be achieved (Hocking 1994). When dry the material requires wetting, and when excessively wet the residual clay and mica content give the fill an apparent thixotropy which eventually disperses when drained. Once compacted the material develops a natural 'set' due to the cementing action of the clay/mica, resulting in a marked increase in stability.

As selected granular fill
Table 5 showed that sands and gravels are acceptable constituents of Class 6 granular fills. China clay sand is therefore potentially suitable for such use and would almost invariably satisfy the requirements for plasticity and chemical composition. Whether or not it satisfied the grading and particle strength requirements would depend on particular circumstances. Figure 10 shows that, by careful selection, china clay sand can be obtained which will satisfy the more stringent requirements for unbound granular sub-bases, and it has been used for this purpose. It therefore follows that it should not be too difficult to find sources of china clay sand which will meet the requirements for selected granular fills. However, selection of suitable materials from a stockpile inevitably increases the price.

As a granular capping material
The requirements for unbound granular capping materials (6F1 and 6F2) as given in the Specification for Highway Works (1998) were summarized in Table 5. This showed that china clay sand is potentially suitable for use in granular capping layers. As it has been used successfully as an unbound granular sub-base material in selected applications, it clearly has a higher potential for use as a granular capping.

As a cement-stabilized capping material
If it does not meet all the grading requirements for an unbound granular capping, china clay sand can be upgraded by stabilization with cement. No problems of excessive plasticity or with chemical composition are likely to arise with china clay sand and it mixes easily with cement. The results in Table 11 show that it can be stabilized with cement to meet the strength requirements for cement-bound sub-bases.

As a granular sub-base material
The Specification for Highway Works (1998) excludes natural sands and gravels from use as Type 1 granular sub-base materials but they are permitted for use as Type 2 sub-base. Figure 10 shows that the coarser varieties of china clay sand meet the grading requirements, given in Fig. 6, for Type 2 materials. The sand is not likely to present any problems with regard to chemical composition, particle strength and durability.

Table 11. *Compressive strength of cement-stabilized china clay sands (Tubey 1978)*

Source of sand	Cement content (%)	7-day strength (MPa)
Melbur	5	2.8
Melbur	10	7.4
Black Alder	5	4.4
Black Alder	10	13.4
Highborough	5	4.8
Highborough	10	11.8

Tubey (1978) showed that china clay was marginally frost-susceptible, but the criteria for frost-susceptibility have since been relaxed and frost-susceptibility is unlikely to be a problem.

As a cement-bound sub-base and base material
The results in Fig. 10 and Table 11 show that china clay sand can be stabilized with cement to produce a material that complies with the requirements for CBM1 and CBM2 cement-bound sub-base material with regard to both grading and strength. The sand would also be potentially suitable for use as the fine aggregate in CBM3 and CBM4 cement-bound road base but might require processing to produce material of the correct grading.

As a concreting sand
China clay sands are used in south-west England as mortar and concreting sands and also for the manufacture of concrete blocks. However, when used in concrete they have a high water demand leading to a reduction in strength which has to be compensated for by increasing the cement content. The concretes are also said to be harsh in texture and to have a poor workability which makes them more difficult to place (BACMI 1991).

Use of stent in road construction
The use of crushed stent as a substitute for primary aggregates in road construction has been encouraged in recent years in Cornwall (Hocking 1994). Using conventional crushing and screening plants it has been possible to produce aggregates that meet the majority of requirements for aggregates given in the Specification for Highway Works (1998). These include drainage filter media, pipe bedding, selected granular fill and Type 1 sub-base. It has also been used to produce coarse aggregates for concrete in accordance with BS 882 (1992).

4. Colliery spoil

Occurrence

Colliery spoil, also known as minestone, is available in the former coal mining areas of the UK (see Fig. 11) in very large quantities. The National Coal Board (later British Coal) placed an emphasis on encouraging the use of minestone as a secondary aggregate and formed the Minestone Executive specifically for this purpose. However, following privatization of the coal industry the Minestone Executive was disbanded and the industry in its present form no longer promotes the use of minestone to the same extent.

Of the wastes and industrial by-products available, colliery spoil was until recently produced in the largest amounts with an estimated annual production of 45 million tonnes (Table 12) available for disposal on land in 1988/89. The collapse of coal mining since then has meant that in the 10-year period from 1988 to 1998 annual production of coal in the UK fell from 102 to 40 million tonnes (see Fig. 12) with a consequent decrease in spoil production to 12.5 million tonnes in 1998, which was only one-third of the amount ten years previously. Even so, colliery spoil is likely to remain the major source of waste available if only because of the estimated 3600 million tonnes available in the stockpile of spoil arising from past production. It is therefore still by far the most significant of the materials potentially available for use in road construction.

The introduction, in the 1950s, of mechanized methods of working increased the proportion of spoil to coal, so spoil production greatly increased between 1950 and 1970 (Table 12). Most of this spoil was tipped on land so that at its peak about 50 million tonnes a year were being tipped.

Composition

Colliery spoil deposits are composed of the waste products from coal mining which are either removed to gain and maintain access to the coal faces or are unavoidably brought out of the pit with the coal and have to be separated at the coal cleaning plant. Wastes from both sources are usually dumped on the same spoil heaps, which are often referred to as 'shale' or 'slag' heaps. These heaps are a prominent feature of former and existing coal mining areas (see Fig. 13). (None of the materials should be regarded as slag, the definition of which, in civil engineering construction, should be confined to materials derived from the extraction of metals from

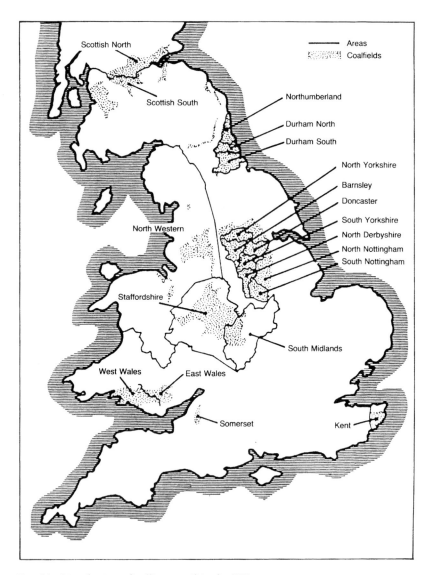

Fig. 11. Distribution of colliery spoil in the UK

their ores (see Chapter 10). Shale is also a misnomer as the tips contain materials other than shale.)

Due to the manner in which they were formed the spoil tips are highly variable in composition. Superimposed is an additional variation arising from combustion in the heaps. When combustion occurs the physical and chemical composition is changed – burnt spoil (burnt shale) differs considerably in its properties from unburnt spoil (minestone). The

Table 12. Ratio of spoil production to coal output in England and Wales 1920–1990 (Whitbread et al. *1991)*

Year	Coal output (million tonnes)	Spoil output (million tonnes)	Spoil output/ coal output (%)
1920	230	9	4
1930	240	12	5
1940	220	18	8
1950	200	15	8
1960	185	38	21
1970	135	55	41
1980	105	50	48
1990	85	45	53

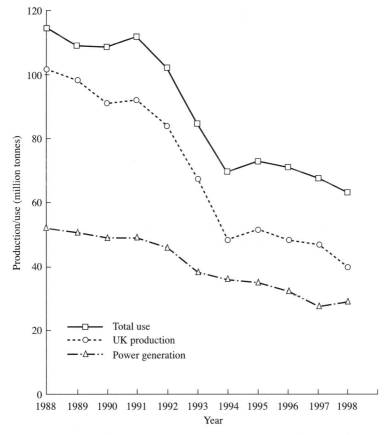

Fig. 12. UK coal production and use 1988–1998 (Annual Abstracts of Statistics 2000)

Fig. 13. Colliery spoil tip

availability of burnt spoil declined following the Aberfan disaster of 1966, as much greater care was taken in the construction of the spoil tips. Taken together with improvements in coal separation this means that the possibility of spontaneous combustion in modern spoil tips is negligible. Burnt spoil is therefore only available from older tips and is also in higher demand than the unburnt material (minestone), because it can be used in road sub-base and base construction. Supplies are therefore being rapidly depleted.

The most common minerals in colliery spoil are quartz, mica and clay minerals and lesser quantities of pyrites and carbonates of calcium, magnesium and iron. Oxidation of the pyrites, which is one of the causes of spontaneous combustion in colliery spoil (see below), can mean that some of the carbonates are converted into their corresponding sulphates. Typical chemical analyses of burnt and unburnt spoil are given in Table 13. The physical and chemical properties of a range of spoils are given in Table 14 and in Figs 14 to 20.

Chemical problems
Spontaneous combustion
Spontaneous combustion is a hypothetical rather than a real problem, but it is a good example of a problem that is unique to a secondary material as it could never arise with the naturally-occurring materials that are normally used in road construction. The problem originates from the presence of coal in unburnt colliery spoil and the possibility of this being ignited by exothermic reactions in the spoil, such as the oxidation

Table 13. Chemical composition of colliery spoil (Sherwood and Ryley 1970, Rainbow 1989)

Component	Burnt spoil (%)	Unburnt spoil (%)
SiO_2	45–60	37–55
Al_2O_3	21–31	17–23
Fe_2O_3	4–13	4–11
CaO	0.5–6	0.4–4.9
MgO	1–3	0.9–3.2
Na_2O	0.2–0.6	0.2–0.8
K_2O	2–3.5	1.6–3.6
SO_3	0.1–5	0.5–2.5
Loss on ignition	2–6	10–40

of pyrites. In the past there has been concern about the use of this material because of the suspected risk, based on observations of burning and burnt-out spoil heaps. This is reflected in the Specification for Highway Works which has a requirement that bulk fill material susceptible to spontaneous combustion should not be used. However, experience over some 40 years has shown that there is no risk because even if combustible

Table 14. Physical and chemical properties of some colliery spoils (Sherwood and Ryley 1970 and unpublished results of British Coal and TRL)

Particle diameter (mm)	Burnt spoils						Unburnt spoils				
	A	B	C	D	E	F	S	T	U	V	W
Particle size distribution (%)											
>40	2	0	6	0	3	0	7	5	6	12	15
20–40	22	14	14	20	18	15	33	25	7	18	15
10–20	26	22	23	23	22	25	26	35	10	15	32
5–10	22	21	22	20	19	17	17	16	15	15	18
2–5	11	12	13	12	12	11	6	6	17	10	8
<2	17	31	22	25	32	32	11	13	45	30	12
Particle density (Mg/m^3)											
	2.65	2.69	2.71	2.72	2.76	2.90	2.60	2.51	– Not determined –		
pH of shale-water suspension											
	6.5	6.8	5.4	4.2	4.5	8.5	– Not determined –				
Soluble sulphate ($g.SO_3$/litre)											
	0.6	1.4	1.6	7.0	6.9	1.5	– Not determined –				

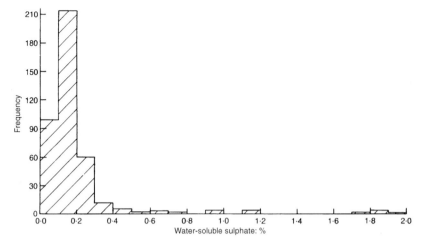

Fig. 14. Range of water-soluble sulphate content values (as SO₃) of unburnt colliery spoils (Rainbow 1989)

material is present the required state of compaction within an embankment is such that the air content is too low to allow combustion. This fact is recognized by the Highways Agency which does allow unburnt colliery spoil to be used for bulk fill.

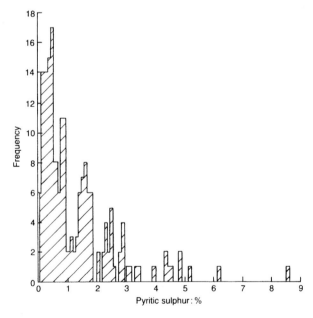

Fig. 15. Range of pyritic sulphur content values (as S) of unburnt colliery spoils (Rainbow 1989)

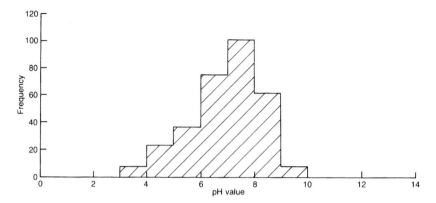

Fig. 16. Range of pH values in unburnt colliery spoils (Rainbow 1989)

Sulphates
The soluble sulphate content of unburnt spoils is generally too low to be a
serious problem (see Fig. 14) but soluble sulphates may occur in large concentrations in burnt colliery spoil due to the oxidation of pyrites in the
unburnt spoil during combustion (Sherwood and Ryley 1970). Sulphates
may cause problems by migrating from the spoil and reacting with the
cement in concrete and other cement-bound materials to form products
that occupy a greater volume than the combined volume of the reactants.
With all forms of sulphate attack, water is an essential part of the reaction.
The water present in the spoil at the time of placing will be insufficient to
dissolve very much sulphate, so unless extra water is able to enter the

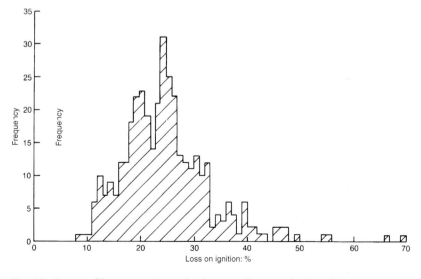

Fig. 17. Range of loss on ignition of unburnt colliery spoils (Rainbow 1989)

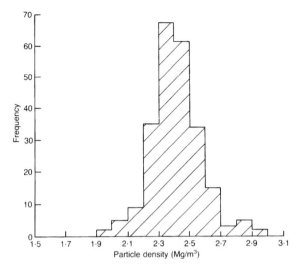

Fig. 18. Range of particle density values of unburnt colliery spoils (Rainbow 1989)

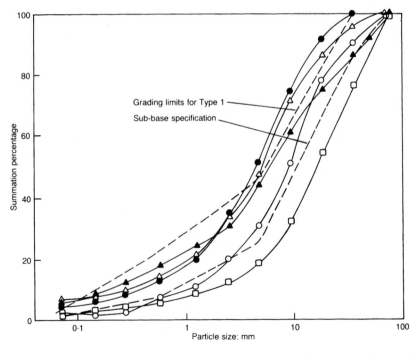

Fig. 19. Particle size distribution of some samples of burnt colliery spoils (Sherwood 1987)

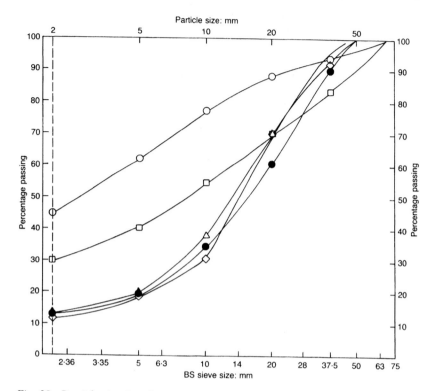

Fig. 20. Particle size distribution of some samples of unburnt colliery spoils (from British Coal, unpublished)

material no appreciable attack will occur even if high concentrations of sulphate are present.

The expansive reaction between sulphates and concrete has been known for at least 100 years and the problem is well understood. It arises because of the ability of sulphates to react, in the presence of water, with the hydrated cement in concrete to form calcium sulphoaluminate (ettringite). This occupies a greater volume than the combined volume of the reactants, which leads to the expansion and disintegration of the concrete. The main reaction is between calcium sulphate ($CaSO_4$, $2H_2O$) and the hydrated calcium aluminates:

$$3CaO.Al_2O_3.19H_2O + 3(CaSO_4.2H_2O) + 6H_2O$$

$$= 3CaO.Al_2O_3.3CaSO_4.31H_2O$$

Magnesium and sodium sulphates behave in a similar manner but are more harmful because of their higher solubilities. Magnesium sulphate is also able to react with the hydrated calcium silicates.

The reactions between sulphates and the hydrated silicates and aluminates lead to products that occupy a greater volume than the combined volume of the reacting constituents. Although the role of these reactions in initiating the expansion is not in question, Lea (1970) pointed out that there are difficulties in ascribing the observed expansion directly to the increased volume of solids. This is because the expansions that occur are much greater than would be expected from this cause alone. He reviewed the possible reasons that have been put forward for this discrepancy, which include osmotic effects and secondary expansion resulting from the destruction of the cementitious materials.

Determination of sulphates. Determination of the total sulphate content of colliery spoil is relatively easy. The method given in BS 1377: Part 3 (1990) for soils may also be used for colliery spoil; it involves extraction of the sulphates in the spoil with dilute acid followed by gravimetric determination as barium sulphate. A similar method, for the determination of the total sulphate content of aggregates, is included in BS EN 1744-1: 1998 and BS 812: Part 118 (1988). This latter standard includes data on the precision, which is quite good, of the test method.

Sulphate attack of concrete can only occur if the sulphates are able to migrate from the spoil and attack concrete in the vicinity of the placed spoil. Determination of the total sulphate content of the spoil is thus of limited value because it gives no indication of the potential for the sulphate to pass into solution. Measurement of the water-soluble sulphate content is therefore the preferred method for determining the degree of risk which sulphates may present.

The usual procedure is to extract the sulphate ions from the spoil with a limited amount of water to restrict the influence of the sparingly soluble calcium sulphate on the result. BS 812: Part 118 (1988), BS 1047 (1983) and BS 1377: Part 3 (1990) all give basically the same method which involves the removal of sulphate ions by shaking one part by mass of the material with two parts by mass of water and expressing the sulphate content (as SO_3) of the aqueous extract in terms of grams per litre (g.SO_3/ litre) (note: the soluble sulphate content is occasionally reported as a percentage, as in Fig. 14, 2 g.SO_3/litre is equivalent to 0.2% of SO_3). The solubility of calcium sulphate is only 1.2 g.SO_3/litre and, if the sulphate ion concentration exceeds this value, other more soluble sulphates must be present. The Specification for Highway Works requires that all fill and unbound sub-base and base materials, 'when placed within 500 mm of cement-bound materials, concrete pavements and concrete products ... shall have a soluble sulphate content not exceeding 1.9 g of sulphate (expressed as SO_3) per litre'.

The sulphate content of soils and aggregates has for very good reasons traditionally been expressed in terms of SO_3 but some recent specifications express sulphate content in terms of SO_4. Whilst to the uninitiated this may seem more logical it is unfortunate that such confusion has arisen and it is therefore important to verify the terms in which sulphate is expressed both in the results of the chemical analysis and in the specification that is being used. Sulphate content expressed as SO_3 can be converted to SO_4 by multiplying by 1.2; the factor for the reverse calculation is 0.83.

Sulphides

Sulphides in the form of iron pyrites (FeS_2) frequently occur in unburnt colliery spoil (see Fig. 15) but are less likely to be present in burnt spoil because they become oxidized to sulphates during combustion. Iron pyrites is of little concern but when exposed to air it may oxidize to jarosite ($KFe_3(SO_4)_2(OH)_6$) by the following sequence of reactions:

$$2FeS_2 + 2H_2O + 7O_2 = 2FeSO_4 + 2H_2SO_4 \qquad (1)$$

$$4FeSO_4 + O_2 + 2H_2SO_4 = 2Fe(SO_4)_3 + 2H_2O \qquad (2)$$

$$3Fe_2(SO_4)_3 + 12H_2O = 2HFe_3(SO_4)2(OH)_6 + 5H_2SO_4 \qquad (3)$$

Cation exchange reaction with potassium minerals in the spoil then leads to the formation of jarosite. If calcium carbonate is present the sulphuric acid released by (1) and (3) will react to form gypsum:

$$CaCO_3 + H_2SO_4 = CaSO_4 + 2H_2O + CO_2 \qquad (4)$$

Most of these reactions are exothermic and the heat produced is one of the reasons for spontaneous combustion in unburnt spoil. As the reactions produce sulphates it follows that they may then attack concrete and other cementitious materials in the manner considered above. However, apart from this, additional expansive reactions may arise which could cause heave in an exposed formation even if no concrete is present. The calculated volume expansion from pyrites to jarosite is reported to be 115%, and the volume expansion from pyrites to ferric sulphate is 170%. The formation of gypsum is accompanied by a large increase in volume and this is believed to be the main cause of expansion in pyritic shales, which has occasionally proved to be a problem when such shales have been used in construction (see Fig. 21) (also Nixon 1978, Hawkins and Pinches 1987). Instances have been reported where heave occurred even though little calcium carbonate was present (Collins 1990).

Determination of sulphides. A method for the determination of the total sulphur content contributed by sulphates and sulphides is included in BS 1047 (1983) and in BS EN 1744-1 (1998) and the method could also be applied to colliery spoil. In this method the sulphides are oxidized to sulphates with a suitable oxidizing agent and the combined sulphate content is determined by the method described previously for sulphates. The value obtained, which is expressed in terms of the total sulphur content, is made up of the initial sulphate content plus the contribution to the sulphate content resulting from the oxidation of the sulphides. The sulphide content of the material may then be readily calculated if a separate determination is also made of the original sulphate content of the material before the oxidation of sulphide. A method for determining the acid soluble sulphide content is also included in BS EN 1744-1 (1998).

Fig. 21. Heave of floor slab as a result of expansive reactions in unburnt colliery spoils

Uses of colliery spoil in road construction

The permitted uses of colliery spoil as given in the current (1988) Specification for Highway Works are summarized in Table 15 and in more detail below.

Bulk fill

Colliery spoil has been successfully used as bulk fill material. At the peak of the motorway building programme in the early 1970s (Fig. 22) it was estimated that about 8 million tonnes per year were being used. This was the largest amount of any waste material or industrial by-product being used in road construction and it represented the biggest single useful commercial outlet for colliery spoil.

The main problem with colliery spoil as a fill material is its variability within a deposit. Spoil heaps may contain burnt spoil, partially-burnt

Table 15. Permitted uses of colliery spoil in road construction

Material	Embankment and fill	Capping	Unbound sub-base	Cement-bound sub-base	Cement-bound road base	PQ concrete	Bitumen-bound layers
Burnt spoil	✓	✓	✓	✓	×	×	×
Unburnt spoil	✓	×	×	✓	×	×	×

Fig. 22. Road embankments of colliery spoil on M62 motorway, Yorkshire 1974

spoil, unburnt spoil and mine tailings, with quantities of all four occurring quite close together. However, except for mine tailings, all are suitable for fill. Visual inspection of the tip should ensure that the type of material delivered to the site does not vary too frequently since control of compaction may be difficult. When considering compaction requirements on the basis of the recommendations given in the Specification for Highway Works, most unburnt spoils are classified as 'well-graded granular and dry cohesive soils'. However, some, while being acceptable for use as fill, may have untypically high fines or moisture contents and could more appropriately be considered as 'cohesive soils'.

Sulphates. Normally the presence of sulphates in bulk fill will not be a problem; it becomes a problem only if the fill is placed close to concrete structures when the limits given above apply.

Frost-susceptibility. Most unburnt spoils are not susceptible to frost when tested by BS 812: Part 124 (1989). On the other hand, burnt spoils are usually highly susceptible to frost. However, frost-susceptibility of the compacted fill is unlikely to be a problem if the fill is more than 450 mm below the finished road surface. In the UK frost penetration rarely exceeds 450 mm, and in all but the most lightly-trafficked roads the fill material will be below this depth. If the fill material is likely to be subjected to frost penetration a check should be made to ensure that it is non-susceptible to frost-heave as defined by the BS test.

As selected granular fill

Table 4 showed that unburnt spoil is excluded from use as a selected granular fill material. This is not surprising as, although it can on occasion seem to be granular, it does not possess the properties generally associated with such materials. The coarse particles are not discrete and are usually aggregations of smaller particles, which means that the long-term stability of the aggregated particles is open to question.

Well-burnt spoil is, however, specified by name for many applications of selected granular fill. Figure 19 shows that burnt spoil can be obtained which meets the grading requirements for granular sub-base material and can therefore easily satisfy the more relaxed grading requirements of selected granular fills.

Burnt spoil is excluded from some uses but the exclusions can be justified. More surprising than the exclusions are the inclusions where, for example, burnt colliery spoil is permitted to be used as fill to Reinforced Earth (6I), the upper bed (Class 6L) and surround (Class M) to steel structures. There is no danger of unsuitable spoils being used as the specification also imposes limits for pH, sulphate and chloride content and redox potential. However, these limits are such as to make it virtually impossible for most spoils to meet the specification.

As selected cohesive fill

The Specification for Highway Works excludes unburnt spoil from use as a selected cohesive fill for all categories except as fill to structures (Class 7A). The omission of unburnt spoil from the materials that are potentially suitable for stabilization with lime (Class 7E) or with cement (Class 7D) is open to question and will be considered further.

The exclusion of unburnt spoil as a fill material to Reinforced Earth (Class 7C) has been critically reviewed by West and O'Reilly (1986). They pointed out that the failure of a Reinforced Earth structure could have serious consequences and it is therefore essential that all precautions should be taken to ensure that the structure will perform satisfactorily over its design life. As there is a small, but significant, risk that this would not be the case if unburnt spoil were to be used as the fill for Reinforced Earth, they endorsed its exclusion from the list of suitable materials. They added that in any case the potential outlet for unburnt spoil as a fill material for Reinforced Earth is infinitesimal in relation to the outlet for its use in bulk fill.

As a granular capping material

The requirements for unbound granular capping materials (6F1 and 6F2) as given in the Specification for Highway Works were listed in Table 5. Unburnt spoil is logically excluded but burnt spoil is permitted for use provided that it meets the requirements of the Specification. No limits for the sulphate content are given but an overriding factor would be the general restriction on the maximum sulphate content of any material placed in proximity to concrete. If the capping were to come within 450 mm of the road surface it would also be necessary to check that the material was not frost-susceptible.

As a stabilized capping material
Burnt spoil could be stabilized with cement to form a stabilized capping material, but generally speaking there would be little reason for doing so as it would be suitable for use in unbound form. Unburnt spoil is excluded by the Specification for Highway Works from use as a stabilized material for capping layers. This can be justified but its exclusion is, at first sight, anomalous because cement-stabilized spoil is permitted to be used for sub-base construction provided that it fulfils all the requirements of Clause 1036 for cement-bound material category 1 (CBM1). It thus does not seem logical to permit it for sub-base while prohibiting it from less onerous use in a capping layer.

Part of the reason lies in the manner in which the clauses for cement-stabilized capping (Clause 614) and CBM1 (Clause 1036) are drafted. Of the two the latter is far preferable because it specifies the properties of the material and the stabilized end-product in considerable detail but does not define the actual material to be used. Hence, there are no restrictions on the material as such but requirements on its properties before and after it has been stabilized. Clause 614, on the other hand, attempts to define the materials to be used but gives inadequate detail as to the property of the stabilized material that is being sought.

There is much evidence to show that unburnt colliery spoil can be successfully stabilized with cement (Kettle and Williams 1978, Tanfield 1978) and a guide on the subject was published by the National Coal Board (1983). Well publicized failures, due to expansion of the cement-stabilized material after compaction, have occurred with its use (*New Civil Engineer* 1980) and this may explain why it has been excluded. However, the testing regime which is specified for cement-stabilized sub-base and base materials has, since 1986, included a durability requirement which should detect whether any problems are likely to occur with cement-stabilized materials. It is for this reason that the Specification for Highway Works does not attempt to define the actual materials that are to be used for CBM1 sub-base materials.

The ability of the durability requirement, given in the Specification, to detect whether problems with the expansion of cement-stabilized spoil were likely to arise has been confirmed by Thomas *et al.* (1987). Carr and Withers (1987) showed that two types of expansion can occur in cement-stabilized spoil: one, in the short-term, is due to hydration of the clay minerals within the spoil; the other, which can occur in the longer-term, is due to sulphate attack of the cement matrix.

It is probable that unburnt spoil could be stabilized with lime because it contains clay minerals that can react with lime. However, as little work has been done on this subject, lime-stabilized unburnt spoil is excluded as a capping material.

As an unbound sub-base material

'Well-burnt non-plastic shale' is one of the materials mentioned by name in the Specification for Highway Works (1998) as being acceptable as a granular sub-base material. The problem lies with the definition of 'well-burnt' as there are large differences in the physical properties of well-burnt and unburnt colliery spoil which can make the former eminently suitable for use as a granular sub-base and the latter totally unsuitable. The main difficulty in using spoil as an unbound granular sub-base material, therefore, lies in distinguishing how well burnt it is and in avoiding materials that are only partially burnt. Colour gives some indication but is not wholly reliable. Some red colliery spoils which appear well-burnt may have only been fired on the outside and may have 'black hearts'. Conversely, some well-burnt spoils may have a black appearance due to their having been fired in a reducing atmosphere, but this does not necessarily imply unsuitability. In cases of doubt, hitting the aggregate with a hammer can distinguish well-burnt from unburnt materials as the former ring when struck compared to the dull sound emitted from unburnt spoil.

Figure 19 shows that burnt colliery spoil may be obtained with a particle size that satisfies the grading requirements for granular sub-bases. However, it is doubtful if it could meet the strength and durability requirements that are now specified. All unbound granular sub-base materials are required to have a minimum soaked 10% fines value (TFV) of 50 kN. It is unlikely that many spoils could meet this requirement, even though in practice they would be suitable. Up to the publication of the 1986 edition of the Specification this was recognized and well-burnt shale was exempted from this requirement.

The imposition of a requirement for a minimum TFV originates from work by Hosking and Tubey (1969). This work was carried out on natural aggregates, most of which pass the TFV requirement with ease. Dawson and Bullen (1991) have shown that the requirement unjustly excludes furnace bottom ash (FBA) for use as an unbound sub-base material (see Chapter 5) and the same is true of burnt colliery spoil.

The more recent editions of the Specification for Highway Works (since 1991) have made it still more difficult for burnt colliery spoil to fulfil the requirements for unbound sub-base materials as there is now a durability requirement in terms of the magnesium soundness value. This is a result of research by Bullas and West (1991) aimed at defining the terms 'clean, hard and durable' that are often used as a subjective description of aggregates to be used in road construction. For aggregates used in bitumen macadam road base, Bullas and West (1991) suggested a magnesium sulphate soundness value of 75, as determined by the method given in BS 812: Part 121, as the criterion for durability. Although this research was specifically aimed at bitumen macadam road base and did not deal with the durability of unbound sub-base materials, the 1991 edition of the specification stipulated the same sulphate soundness value of 75 for unbound granular materials.

The reasons for imposing more stringent requirements on burnt spoil are not given, and for minor roadworks there would seem to be no

Table 16. Effect of the addition of cement on the frost-heave of burnt colliery spoil (Fraser and Lake 1967)

Sample No.	Frost-heave (mm)	
	Without cement	With 5% cement
1	22	7
2	41	13
3	45	11
4	30	8
5	6	0

reason why the specification in use up to 1986 should not continue to be used. This would dispense with the need to determine TFV and soundness.

Burnt colliery spoil frequently contains sulphates in sufficient quantity to preclude its use as a sub-base in situations where it is within 500 mm of concrete structures or road pavement layers containing cement. It is also usually highly susceptible to frost, which means that it should not be used within 450 mm of the road surface. However, Table 16 shows that the frost-susceptibility may be reduced by the addition of cement.

As a cement-bound sub-base material
Provided they can meet all the relevant requirements, the Specification for Highway Works permits the use of both unburnt and burnt colliery spoil as the 'aggregate' in CBM1 and CBM2. As mentioned above this gives rise to a slight anomaly in that cement-stabilized unburnt spoil is not permitted to be used as a capping layer.

The presence of high sulphate contents may occasionally cause problems. There is no requirement in the specifications for CBM1 and CBM2 for the sulphate content to be determined as it is assumed that problems arising from the presence of sulphates would be demonstrated by failure of the test specimens to meet the soaking requirement given in Fig. 8.

Other uses
There is very little potential for colliery spoil to be used above sub-base level. Burnt spoil might meet the grading and strength requirements for CBM3 and CBM4 cement-bound road base material given in Fig. 8 but it is specifically excluded from use.

5. Construction and demolition wastes

Occurrence

Construction and demolition wastes are materials such as concrete, masonry, bituminous road materials, etc., arising from the demolition of buildings, airfield runways and roads (Fig. 23). In addition to these 'hard materials', which are eminently suitable for recycling, a large quantity of soil is produced which has no potential for use as a secondary aggregate and a third category also exists where the hard materials and soils are inextricably mixed.

A survey of the materials available in England and Wales (DETR 2001a) showed that 72.5 million tonnes of construction and demolition wastes (excluding road planings) were produced in 2000, made up of:

  39 million tonnes (46%) of hard demolition waste such as concrete and bricks
  24 million tonnes (33%) of soil
  15 million tonnes of mixed hard materials and soil plus minor amounts of other inert materials.

This survey showed that, of the 39 million tonnes of hard waste produced, 23 million tonnes were recycled to produce secondary aggregates amounting to 31% of the total waste stream. Another 18% was considered to have some potential for recycling as aggregate and the remaining 51% had limited potential. The proportion of construction and demolition waste that was recycled (31%) in 2000 represents an enormous advance on a 1994 estimate (Howard Humphreys 1994) which found that only 4% was recycled.

Construction and demolition wastes which are then re-used as substitutes for natural aggregates clearly fall into the category of recycled aggregate (RCA) because they originally contained natural aggregates that are being recycled. This is the definition used by the *BRE Digest* (BRE 1998) on recycled aggregates, which defines them as 'crushed concrete and brick masonry'. However, the position is clouded by the frequent use of the term also to refer to secondary aggregates, produced from the various processes referred to elsewhere in this book. Thus a Press Release of the Quarry Products Association (QPA 2001) is entitled 'Recycled Aggregates' when it is quite clear from the context that it includes materials such as colliery spoil, slate waste, china clay sand, etc. which are more properly referred to as secondary aggregates.

Fig. 23. Removal of concrete road pavement from M4 motorway

The recycling of construction materials has long been recognized to have the potential to conserve natural resources and to reduce the energy used in production. In some countries it is a standard alternative for both construction and maintenance, particularly where there is a shortage of aggregate. In the UK a plentiful supply of good-quality aggregate, relatively short distances between quarry, mixing plant and site and a wide range of specifications have hitherto reduced the need for recycling.

However, the climate of opinion is rapidly changing because of increasing concern to conserve natural resources and reduce the environmental impact of road construction. As a result, a considerable amount of research has been carried out recently into methods of re-using construction materials, particularly those arising from the reconstruction of roads. This work has led to the inclusion since 1991 in the Specification for Highway Works of the option of using recycled materials as a replacement, either in whole or in part, for natural aggregates.

Composition

Apart from asphalt road planings, Mulheron (1991) distinguished four main categories of construction and demolition wastes. These are, in order of potential use:

(*a*) clean crushed concrete: crushed and graded concrete containing less than 5% of brick or other stony material

(*b*) clean crushed brick: crushed and graded brick containing less than 5% of other material such as concrete or natural stone

(*c*) clean demolition debris: crushed and graded concrete and brick

(*d*) crushed demolition debris: mixed crushed concrete and brick that has been screened and sorted to remove excessive contamination, but still containing a proportion of wood, glass or other impurities.

Table 17. Classes of recycled aggregate (RCA) (BRE 1998)

Class	Origin	Brick content (%)
RCA(I)	Brickwork	0–100
RCA(II)	Concrete	0–10
RCA(III)	Concrete & brick	0–50

Table 17, taken from the *BRE Digest*, gives three classes of RCA depending on the relative amounts of brick and concrete. The three categories in Table 17 are similar to the first three categories of Mulheron (1991). The *Digest* embraces Mulheron's fourth category by giving limits for the impurities (glass, wood, metals, sulphates, etc.) that can be accepted for any particular application.

Crushed concrete
Any crushed aggregate produced from construction and demolition wastes tends to be referred to as crushed concrete (O'Mahoney 1990), but the term should be reserved for crushed concrete produced from the break-up and crushing of concrete slabs from road and airfield pavements. Concrete is also available from the demolition of buildings but it is likely to be reinforced, which makes it difficult to crush, and contaminated with other building materials.

Crushed concrete arising from the demolition of disused airfield runways was widely available after World War II. This source gradually declined but may temporarily regain importance as, following the end of the Cold War, further airfields will probably be closed. Concrete from this source and from pavement-quality concrete removed from roads can, provided it does not contain reinforcement, be crushed to produce an aggregate which can be assumed not to contain any harmful components. Any material containing other constituents as well as crushed concrete, e.g. brick, glass, asphalt, etc., would, under Mulheron's classification (*a*)–(*d*) above, go into the category of crushed demolition debris.

Unless the source is known there is unfortunately no generally recognized method of classifying crushed concrete so that it can be readily distinguished from other construction and demolition wastes. A Dutch classification for recycled granular base materials described by Sweere (1991) defined crushed concrete in the following manner.

(*a*)   *Main components*: at least 80% by weight of crushed gravel or crushed aggregate concrete; at most 10% by weight of other broken stony material, the particles of which shall have a particle density of at least 2.1 Mg/m^3.

(*b*)   *Additional elements*: at most 10% by weight of other crushed stone or stony material. As for broken asphalt, this shall not exceed 5%.

(c)   *Impurities*: at most 1% non-stony material (plastic, plaster, rubber, etc.); at most 0.1% decomposable organic matter such as wood and vegetable remains.

Recycling of crushed concrete.   The Specification for Highway Works allows the use of crushed concrete as a substitute for natural aggregates for most purposes. However, although not specifically stated, it can be assumed that, where it is allowed, the Specification envisages it will relate to crushed concrete as defined above and not to demolition debris. This is made particularly apparent in those instances where the quality of the crushed concrete is critical to the performance of the material to be produced from it. Thus, when crushed concrete is to be used as the aggegate in cement-bound road base materials CBM3 and CBM4 and for pavement or for pavement-quality concrete, the specification requires that the crushed concrete shall comply with the grading requirements of BS 882 (1992).

Concrete made from recycled aggregates derived from crushed concrete will inevitably contain more hardened cement paste than concrete made from other aggregates. This means that the elastic modulus will be reduced and shrinkage, creep and the coefficient of thermal expansion will all be increased compared with similar strength concretes.

Most recycled aggregate will in all probability be derived at least in part from crushed concrete and a specification for concrete from recycled aggregates has been published by RILEM (1994). This classifies coarse recycled aggregates into the following three categories.

Type 1 aggregates, which are implicitly understood to originate from masonry rubble.

Type 2 aggregates, which are implicitly understood to originate from concrete rubble.

Type 3 aggregates, which are implicitly understood to consist of a blend of recycled aggregates and natural aggregates.

Recycling crushed cement-bound sub-base and road base.   Some design procedures envisage a cement-bound road base which at the end of its useful life is removed and replaced. For whatever reason, the removal of long sections of cement-bound road base or sub-base can give rise to large quantities of material which may have potential for re-use. For example, they may be crushed to provide a good-quality granular material which meets all the requirements for granular sub-base materials (Figs 24 and 25).

The granular material, produced by crushing, may also be re-stabilized with cement to produce a cement-stabilized material in a lower category. Figure 26 shows the relation between cement content and strength for a crushed CBM3 road base. This shows that it can easily be made to meet the requirements for CBM2 sub-base and that the material may have residual cementitious properties as a result of the exposure of unhydrated cement during crushing.

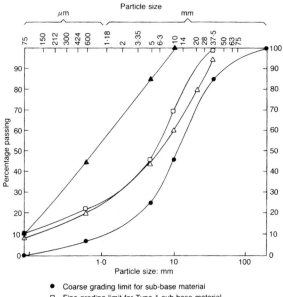

● Coarse grading limit for sub-base material
□ Fine grading limit for Type 1 sub-base material
▲ Fine grading limit for Type 2 sub-base material
△ Grading of crushed CBM3 road base

Fig. 24. Grading curves of crushed CBM3 road base (Sherwood 1993)

Fig. 25. Crushed CBM3 road base for use as Type 2 granular sub-base

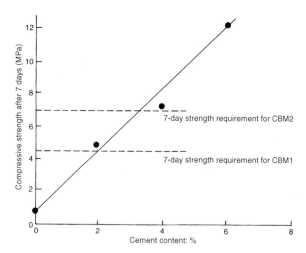

Fig. 26. Relation between cement content and strength for crushed CBM3 road base material (Sherwood 1993)

Crushed brick

The fate of demolished brickwork depends both on the type of brick and the type of mortar. It is often economic to recover the bricks, for which there is a significant second-hand demand, but cheaper bricks such as flettons would usually be sold as hardcore. The type of mortar used will influence the decision; lime mortar is easily separated from the bricks but cement-containing mortars are very difficult to remove. Contamination from either lime or cement mortar is not likely to be a problem but contamination from gypsum plaster could result in the crushed bricks having unacceptably high sulphate contents. Even without contamination from gypsum some bricks have soluble sulphate contents high enough to be deleterious. Where bricks are available in quantity but are unsuitable for re-use they may be crushed to provide an excellent granular sub-base (Fig. 27).

Crushed demolition debris

Crushed demolition debris other than crushed concrete and crushed brick has no applications for use except as general bulk fill. Provided that it meets the requirements given earlier there is no reason why it should not be used for bulk fill, but care needs to be taken with its use because it can be very heterogeneous. Rubble containing timber should be avoided because, when it rots, cavities will be left in the fill.

Recycled road pavement materials

Waste from road reconstruction forms only one quarter of the total amount of 'hard' construction waste produced. However, it clearly has a high potential for re-use in roadmaking as not only does it occur on site, but it is also likely to retain at least some of the properties that

Fig. 27. Crushed brick as a granular sub-base material

originally made it suitable for use in road construction. It is therefore considered separately from other construction and demolition waste in this chapter.

New road construction may incorporate any of the materials considered elsewhere in this book. However, this section is specifically concerned with the recycling of existing materials arising from the re-construction of a road pavement. It is primarily, although not exclusively, concerned with recycling of the materials within the road itself. The removal of surplus material for use in another road is more akin to the use of construction and demolition waste considered earlier.

Materials removed from a road have obvious potential for recycling because by definition they are composed of the same materials that are traditionally used in road construction. They may have ceased to function for their original purpose but they can be upgraded or used for a different application where less stringent requirements are acceptable. Earland and Milton (1999) define this as using the existing highway as a 'linear quarry' from which roadstone aggregates can be reclaimed. Consequential environmental benefits from which the general public, central and local government and the transport industry all benefit are:

- a reduction in the extraction of primary aggregates
- a reduction in the amount of construction traffic servicing maintenance sites
- a reduction in the use of energy
- a reduction in traffic congestion from shorter duration
- the promotion of sustainable development and improvement of the road network.

Reclaimed bituminous pavements.   About 95% of the material is the aggregate from which it was made with the remainder comprising the bitumen binder. Milling and crushing yields a granular material consisting of the coated aggregate. If stockpiled for excessive periods (particularly in warm weather) the particles may stick together but they can be readily broken up. Some degradation of the particles inevitably occurs during removal from the road. The material has a high potential for recycling and an OECD survey (1997) showed that among EU member countries at least 75% was re-used – in the UK this amounted to 90%. Cornelius and Edwards (1991) have shown that road planings can be recycled to produce material which can comply with the relevant specifications when it is produced using modified drum-mixers and laid using conventional plant.

Reclaimed concrete pavements.   Crushed concrete, arising from the demolition of a concrete road pavement or airfield runway consists of high-quality well-graded aggregates bonded by hardened cement paste. It may be contaminated with chloride ions from the use of de-icing salt and with sulphates from contact with sulphate-rich soils. The material is highly angular with a rough surface texture and has a lower particle density and higher water absorption than the original aggregate from which it was derived. In the UK about 75% of the material is recycled.

Reclaimed base and sub-base materials.   Given the wide range of materials that can be used in the construction of bases and sub-bases this is a very heterogeneous group of materials about which it is difficult to generalize. They are primarily used for cement-stabilized base, unbound base and fill material. The OECD survey (1997) showed that, for most countries, the amount of recycling was low with the UK recording a figure of only 5%.

Mixed materials.   The three classes of material already mentioned may also occur as a mixture when little care is taken with the removal of the individual layers of the road pavement. Because of the heterogeneous nature there is less scope for recycling and in the UK only 5% was recorded (OECD 1997) as being re-used.

Recycling of bituminous materials

Methods and equipment for recycling old bituminous pavements into new pavements are well developed and widely used. According to a review by the OECD (1977) the benefits of such recycling are:

- reduced transport costs and fuel requirements
- reduction in aggregate requirements and elimination of potential disposal problems
- up to 75% reduction in bitumen content required
- reduction in fuel required for drying (see Table 18)
- generally lower emissions during construction.

The savings in energy can be considerable. Table 18 shows the estimated energy consumption on a road project in Kent in the production of raw

Table 18. Energy consumption in the production of raw materials from asphalt (Hubert 1987)

Material	0% recycle (MJ/tonne[a])	40% recycle (MJ/tonne[a])	80% recycle (MJ/tonne[a])
Sand	9.1	3.6	1.2
Crushed aggregate	27.7	20.0	15.6
Bitumen	75.1	47.3	27.8
Planings	0.0	8.5	12.7
Total	111.9	79.4	57.3
Energy savings (%)	0.0	29.0	48.8

[a] Per tonne of asphalt produced

materials for central recycling of hot-rolled asphalt using different percentages of reclaimed bituminous planings.

Recycling methods. (Note: part of the text of this section on recycling methods is largely based on a paper by Brian Hicks (1999) given at a Conference at the Transport Research Laboratory, Crowthorne in December 1999.)

The recycling of bituminous pavement materials can be broken down into two broad categories: hot-mix (see Fig. 28) and cold-mix (see Fig. 29) as follows.

Hot-mix. Typically known as Repave, the hot-mix processes involve heating the surface layer of the road, scarifying it and then reinforcing it with a thin overlay of asphalt. Generally the total depth of treatment is about

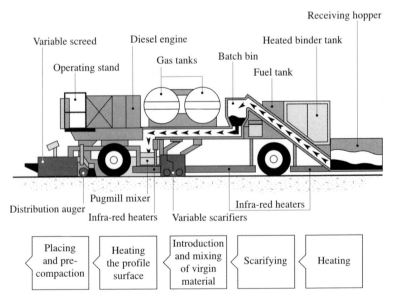

Fig. 28. Diagrammatic representation of hot recycling

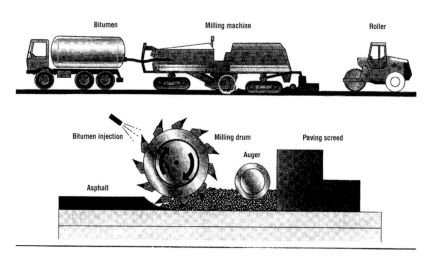

Fig. 29. Diagrammatic representation of cold-mix recycling (from OECD 1997)

50 mm, with cost savings of between 15–20% achieved when compared with more traditional methods. The size of plant involved is considerable and tends to restrict the process to major roads. However, when it can be used, the speed of operation is a bonus. Traditionally the process has used hot-rolled asphalt and precoated chips for the new wearing course.

Cold-mix. Cold-mix in-situ recycling can be carried out to a variety of depths. Shallow cold-mix in-situ recycling, or 'Retread' as it is commonly known, recycles the road to a depth of 75 mm. The process has been in service for over 50 years in this country.

Retread involves firstly scarifying and reshaping the existing road or footway surface. Once completed, virgin aggregate may be added to re-profile the road surface if necessary, alternatively excess aggregate may be removed. After the desired profile has been achieved, bitumen emulsion is applied via a spray tanker and is then harrowed in. This is followed by compaction and a first sealing dressing of 14 mm chippings to close any surface voids. Finally a surface dressing is applied, typically using 6 mm chippings to give adequate texture depth to the surface.

The Retread process has been shown to be a cost effective alternative to planing out and adding a new overlay, and it is claimed to give a cost saving of between 25–35% when compared with conventional reconstruction methods. It is most appropriate for the rejuvenation or reshaping of residential and generally lightly trafficked roads.

The Linear Quarry Project, which started in 1996, aimed to produce a structural design method for cold in-place recycling (CIPR) of bituminous pavements. The study showed that:

- CIPR was 15% less expensive than traditional methods
- both the bitumen and cement-bound recycled treatments appeared to be equal in performance to fresh materials

- more technical control was required in the CIPR process
- the main material variables were water content, quantity and morphology of the fines, binder content and uniformity of mixing.

The work led to the production of a design guide and specification for structural maintenance of highways by cold in-situ recycling (Milton and Earland 1999). A flow chart from this guide is reproduced below as Fig. 30.

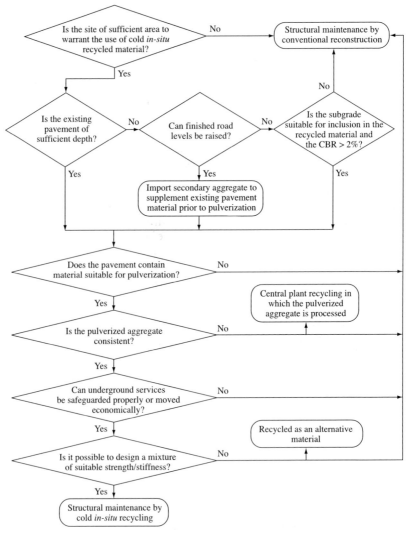

Fig. 30. Flow chart for site evaluation of cold in-situ recycling (Milton and Earland 1999)

For more seriously damaged pavements, *deep cold-mix in-situ recycling* should be considered. As the name implies, this form of recycling treats the road to a far greater depth than does Retread, typically to 300 mm. The process can therefore compete directly with conventional road reconstruction methods. However, like Retread, it still utilizes a cold bitumen system and/or cement for recycling.

The deep cold-mix in-situ process provides two options. First of all as a full width process and secondly as a haunch repair process. Typically, full width recycling operates to between 125 and 300 mm in depth whilst haunch recycling works from 150 to 250 mm. The reduced depth for haunch repair results from the difficulty of compaction, especially if only one metre is recycled, when the depth is typically restricted to around 200 mm.

A combination of haunch recycling and carriageway retread has proved to be an effective solution to a problem which affects many rural roads: surface potholes and deteriorating edges. The road haunch is stabilized with cement to a depth of 200 mm and a width of one metre wide. The full width of the road is then given the retread treatment to a depth of 80 mm.

Specifications for recycled bituminous materials.   The 1991 edition of the Specification for Highway Works recognized for the first time that road planings had a high potential for re-use and it introduced a new clause (Clause 902) which specified that 'up to 10 percent of materials derived from existing carriageway may, with the approval of the engineer, be used in the production of bituminous materials'. This was in marked contrast to the 1986 edition which specified that 'Surplus materials arising from the planing off process shall be disposed of by the Contractor'.

The change in emphasis was continued in 1995 when the Highways Agency published an Advice Note on the 'Conservation and the Use of Reclaimed Materials in Road Construction and Maintenance'. This drew the attention of those responsible for the design, specification, construction and maintenance of roads to 'the opportunities to conserve or to re-use materials arising from road works and the potential uses of reclaimed or re-processed materials from other sources that may be proposed when cost effective'.

One particular outlet for recycled materials was considered to be the repair of road edges of relatively minor roads caused by the overriding of vehicles due to inadequate road width. A study was therefore commissioned to evaluate the use of recycled materials and secondary aggregates and to develop a design guide which was published in 1996 (Potter 1996). The purpose of this guide is to provide engineers with a reference source to construct road edges by using recycled materials. The work leading up to the development of the guide has been summarized by Biczysko (1999). This guide includes a flow chart reproduced below as Fig. 31 which, although designed for haunch repairs, gives good general advice on the assessment of alternative materials for use in road construction.

Further developments occurred in 1998 when Clause 902 was re-written to specify that 'Reclaimed bituminous materials may be used in the production of bituminous road base, basecourse and wearing course. The maximum

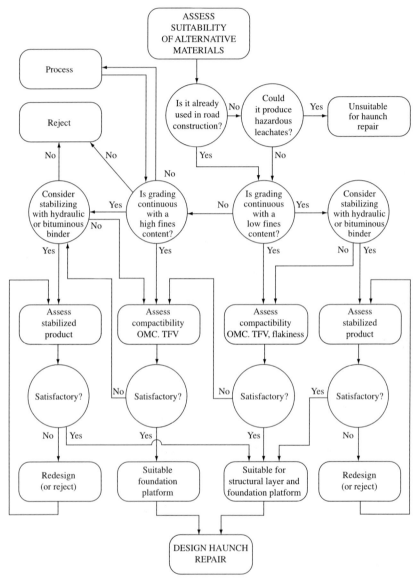

Fig. 31. Decision chart for assessing suitability of alternative materials (Biczysko 1999)

amount of reclaimed materials permitted shall be 10 percent for hot-rolled asphalt and 30 percent for coated macadam basecourse and road base and hot-rolled asphalt basecourse and road base complying with the relevant British Standards'. A new clause (Clause 926) 'In Situ Recycling – The Remix and Repave Process' was introduced at the same time.

6. Glass waste

Occurrence

The principal raw materials for glass manufacture are quartz sand, sodium carbonate and limestone with minor amounts of other minerals depending on the end use. Common soda glass has the approximate composition $Na_2O.CaO.5SiO_2$ and is in fact a super-cooled liquid which behaves as a non-crystalline solid. Glass is of course widely used in building but the principal source of waste glass considered in this section is that from glass containers. UK glass container manufacturers produce about 2 million tonnes/year of glass, the majority of which is clear glass; of this about 0.5 million tonnes is exported as filled bottles (whisky, etc.) but this is offset by the import of 0.75 million tonnes of filled glass containers much of which is beer and wine in green bottles. The net annual consumption in the UK is therefore just over 2 million tonnes of which 0.6 million tonnes is clear glass and 1.6 million tonnes is coloured (mostly green) glass (DETR 2001b).

There are two sources of waste glass: one derives from glass factories and the other from household refuse. Theoretically there is a third source arising from the demolition of buildings but this is so diverse and recovery so difficult that it need not be considered further. The best use to be made of the considerable quantities of waste arising in glass factories is simply to recycle it within the factory so that household waste glass is the only source that needs to be considered. Glass forms about 10% of municipal waste (Fig. 32) but even when the waste is incinerated it would be difficult and totally uneconomic to attempt its recovery. However, in recent years serious attempts have been made to separate waste glass from other household wastes by encouraging disposal through 'bottle banks' and to have separate banks for clear, green and brown bottles.

A fair proportion (22% in 1998) of glass containers (bottles and jars) are recycled, most of which are clear and brown bottles and jars. However, this still leaves well over a million tonnes/year, most of which is green glass derived from imported bottles. This is a significant amount but unfortunately whilst there are many sources of waste glass they are widely scattered. The quantities available at any given locality are therefore relatively small and the cost of transport from source to site militates against its use on a wide scale. This is borne out by an OECD survey (1977) which concluded that waste glass was a material that from an economic standpoint does not seem to be suited for full use in road construction; this view was repeated in a later (1997) OECD report.

Properties

Waste glass can be crushed to produce a product with a high proportion of flat or longish elements. Some of the physical properties of the material are summarized in Table 19. Chemically waste glass from containers is inert but there is inevitably some contamination arising from the normal level of impurities associated with bottle-bank recycling. These impurities are mainly clothing, footware, corks, bottle tops and plastic carrier bags, labels and other impurities deposited with the bottles and jars. These do not necessarily detract from its performance but can be a visible reminder of contamination.

Uses in road construction

Blewett and Woodward (2000) concluded that crushed waste glass has some potential for use as a fill or drainage material but there are no reports of it having been used for these purposes. There are also reports of it having some potential for use as a concrete aggregate, but its use for this purpose should be treated with great caution because of the possibility of the glass aggregate reacting with the alkalies in the cement. For example, a Concrete Society Technical Report (1998) states that 'Aggregates containing detectable opal or comprising man-made glass are likely to be particularly reactive ... and such aggregates are unsuitable for use in concrete'.

The major, if limited, use of crushed glass therefore lies in its use as an aggregate in bituminous mixtures. In the USA many highway agencies routinely allow crushed glass to be used for this purpose where the resulting mixture is known as 'Glassphalt'. Recent research (Nicholls 2000) in Britain has shown that macadam can be manufactured with 30% of the aggregate being replaced by crushed glass without any detrimental effect on the properties of the mixture.

Table 19. Properties of waste crushed glass (Blewett and Woodward 2000)

	Glass	Gravel[a] (for comparison)
Minimum dry density (Mg/m^3)	1.34	1.38
Maximum dry density (Mg/m^3)	2.05	1.72
Particle density (g/cm^3)	2.59	2.75
D_{10} (mm)	2.80	3.50
D_{60} (mm)	5.00	5.00
Coefficient of uniformity	1.78	1.42
Aggregate crushing value	1.7	1.92

[a] 6 mm quartz dolerite gravel

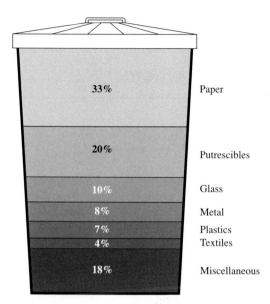

33%   Paper

20%   Putrescibles

10%   Glass

8%   Metal

7%   Plastics

4%   Textiles

18%   Miscellaneous

Fig. 32. Composition of municipal wastes (DETR website 1999)

7. Municipal waste

Occurrence

The waste products of households and commercial premises that are collected by local authorities are known collectively as municipal waste. In 1998/99 a total of 28 million tonnes of municipal waste was produced in England and Wales. The average composition of this waste is given in Fig. 32 on p. 73.

This waste in its raw form has no potential for use in road construction and in 1998/99, 82.4% of it was disposed to landfill and only about 8% was incinerated. Table 20 compares the position among some of the major countries of the EU.

The 28 million tonnes/year of domestic waste produced by the UK is growing at about 3% a year and is on course to double in less than 25 years. Table 20 shows that the UK depends on landfill as a means of disposal of its municipal wastes more than any other country within the EU. Since the imposition of the landfill tax in 1996 the amount disposed to landfill has slightly decreased (from 85% in 1997/98 to 83% in 1998/99) but under the terms of the EU Directive it will have to be drastically decreased. This directive requires that, within five years of its implementation, the weight of municipal wastes disposed to landfill will have to be reduced to 75% of the 1995 figure, to 50% of this figure within eight years and to 35% within 15 years. These are drastic reductions which will be difficult to realize; some reduction can be brought about by an increase in recycling the wastes but it is doubtful if the EU Directive can be achievable without incinerating much of the waste at present going to landfill. Direct incineration results in a 60% reduction in weight with a 90% reduction in volume. Half of the incinerated refuse is ferrous metal which can be magnetically extracted and recycled, while the other half is ash which has some potential for use in road construction.

In 1998/99 only 8% of municipal waste was incinerated within the UK and despite fears of air pollution from waste incineration plants the number of plants and hence the ash produced as a result of incineration will almost certainly greatly increase. The Government has predicted (evidence to Environment Committee of the House of Commons 2001) that if incineration were used on a large scale there would be a need for up to 130 incineration plants with a capacity of 200 000 tonnes or 94 incinerators with a capacity of 250 000 tonnes. This would mean that the annual

Table 20. European municipal waste: total production and percentage utilization

Country	Total[a]	Landfill (%)	Incinerated (%)	Recycled (%)	Composted (%)
UK	28.0	83	8	8	1
Italy	27.0	80	7	3	10
Spain	17.2	74	6	3	17
France	25.8	49	39	6	6
Germany	50.0	34	18	38	10
Netherlands	8.1	12	42	39	7

[a] Million tonnes

amount of incinerated ash produced would rise to about 10 million tonnes, which is about ten times the amount currently produced.

According to the Royal Commission on Environmental Pollution (1993) incineration is an efficient way of recovering the energy present in household wastes in order, for example, to generate electricity. It can help significantly in countering the greenhouse effect because it prevents the wastes from decaying and producing methane gas. Compared with untreated household wastes, the residues left after incineration take up less space and are much more stable. Incineration of household wastes is therefore environmentally preferable to tipping them into a landfill site. The Royal Commission on Environmental Pollution concluded that in the short-term the proportion of waste that is incinerated may well decrease because existing plants cannot meet new standards for emissions. However, it concluded that, on environmental grounds, incineration of domestic refuse was preferable to its use for landfill and 'it

Table 21. Locations and capacity of municipal waste incineration plants in the UK (Source: Sustainable Energy from Waste, DETR/DTI website 1999)

Location	Capacity (000 tonnes/year)	Ash production[a] (000 tonnes/year)
Edmonton (N. London)	600	240
Deptford (S.E. London)	420	168
Birmingham	350	140
Cleveland	220	88
Coventry	220	88
Stoke	200	80
Bolton	150	60
Nottingham	150	60
Sheffield	135	54
Dundee	120	48
Wolverhampton	105	42
Dudley	90	36
Lerwick	26	10.4

[a] Estimated by assuming 60% reduction in weight

offered the best practical environmental option for disposing of the increasing level of household waste'. However, the Environment, Transport and Regional Affairs Committee of the House of Commons argued that incineration was only the second worst option of waste disposal and that far more emphasis should be placed on recycling (House of Commons 2001).

In 1976, 33 incineration plants were operational in England (Roe 1976) which produced between them 1 million tonnes per year of ash. Many of these have since been closed because they did not meet the emission limits set for air pollution and/or because no useful energy was recovered from

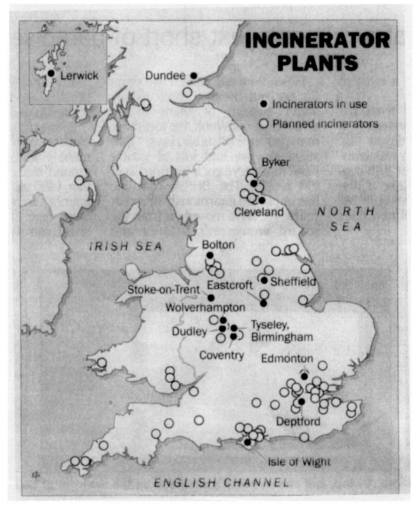

Fig. 33. Map of existing and planned incinerator plants in the UK

them. Table 21 shows the plants that are currently (2001) in operation and their throughput of waste – only those at Edmonton and Nottingham were on Roe's 1976 list. However, as mentioned above, many more are planned and some are already (2001) in the process of being built (see Fig. 33).

Composition

Two types of ash are produced as a result of the incineration of municipal wastes – fly ash, which is taken from the filters in the flues, and bottom ash which is left after the combustion of the materials. The fly ash (which should on no account be confused with pulverized fuel ash (PFA)) has high concentrations of toxic materials which make it entirely unsuitable for use (Bond 2000) and only the bottom ash, which is considered here, has any potential uses in road construction. The incinerated residues consist mainly of clinker, glass, ceramics, metal and unburnt matter. The metal is largely in the form of tin cans and most plants magnetically extract the ferrous metal before discharging the residues. The unburnt matter may be paper, rag or putrescibles, the amounts depending on various factors including the type of furnace, the temperature of the firebed or the length of time the materials spend in it, and on the composition of the raw refuse itself. A complete chemical analysis of the Edmonton ash is given in Table 22.

Uses of incinerated refuse in road construction

Although some ashes may have gradings that make them potentially suitable as a selected granular fill or unbound sub-base material (see Fig. 34), the main use of incinerated refuse, as it comes from the incinerator, is for

Table 22. Chemical composition of the inorganic constituents of ash from the Edmonton incinerator (Roe 1976)

Component	Percentage present
SiO_2	35
Fe_2O_3	26
Al_2O_3	23
CaO	6.3
Na_2O	2.5
ZnO	1.2
SO_3	1.0
MgO	0.7
TiO_2	0.6
MnO	0.3
Other	3.2
Loss on ignition	2.6
Soluble sulphate	2.32 (as $g.SO_3$/litre)
Total sulphate	0.60 (as % SO_3)
pH value	9.4

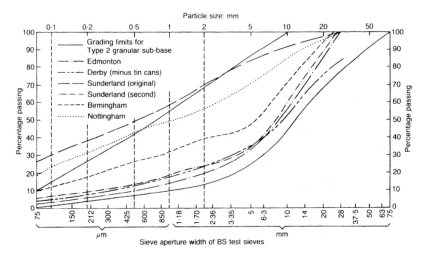

Fig. 34. Comparison of particle size distribution of incinerated refuse from various sources (Roe 1976)

bulk fill. However, careful selection and grading of the ash from the plants at Edmonton, Birmingham and Cleveland means that a product can be obtained that has reasonably consistent properties and which can be used not only as an unbound granular material but also as an aggregate in cement- and bitumen-bound materials. Typical properties of this material which is known as Processed Incinerator Bottom Ash (IBA) are given in Table 23.

Roe (1976) and OECD (1977) cautioned against the use of incinerated refuse as a cement-bound material because it may contain aluminium, heavy metals and glass, all of which are capable of reacting adversely with cement. However, IBA has been used successfully as the aggregate as a CBM3 road base material (Fig. 35).

Table 23. Typical properties of Processed Incinerator Bottom Ash (IBA) (source Ballast Phoenix 2001)

Soaked 10% fines value (TFV)	75 kN
Flakiness Index	21
Acid soluble sulphate (as SO_3)	0.53%
Acid soluble chloride (as Cl)	0.04%
pH value	9–11
Particle density	
Oven dry	2.19 Mg/m^3
SSD	2.35 Mg/m^3
Apparent	2.62 Mg/m^3
Water absorption	7.40%

Fig. 35. CBM3 produced from IBA aggregate being laid as the base for a lorry park (photo courtesy of Ballast Phoenix)

8. Power station wastes (pulverized fuel ash and furnace bottom ash)

Occurrence

About 5 million tonnes per year of pulverized fuel ash (PFA) and 1.5 million tonnes per year of furnace bottom ash (FBA) are currently produced in the UK. The amount produced has declined with the run-down of coal-fired power stations and the increasing proportion of electricity produced by gas-fired and nuclear power stations. However, the decline has not been as drastic as the decline in UK coal production because more coal is now imported. Figure 12 showed that between 1988 and 1998 coal production declined by 60% but the amount of coal used for power production fell by only 10%.

The policy of the electricity generating companies has been to site their power stations as close to the coalfields as is practicable. This means that most of the larger modern power stations are in central England (Fig. 36) and this area produces more than half of the available supplies.

Coal-burning power stations use coal which has been pulverized to a fine powder. When the pulverized coal is burned in a furnace at the power station it produces a very fine ash which is carried out of the furnace with the flue gases. This ash is PFA and accounts for about 75–85% of the ash formed from the burnt coal. The remaining, coarser fraction of the ash falls to the bottom of the furnace where it sinters to form the FBA.

Confusingly, PFA is also known as fly ash and is described as such in European Standards and in many countries. However, the term fly ash is also used to describe ashes produced from furnaces other than those from coal-burning power generation, e.g. from the incineration of municipal waste (see Chapter 7). Such ashes will be entirely different in their chemical and physical properties and it is important that they should not be confused with PFA.

PFA is removed from the flue gases by mechanical and electrostatic precipitation and is initially collected in hoppers; it can be supplied as dry powder (hopper ash) in this state. When required this ash can be passed through a mixer-conveyor plant where a measured amount of water can be added. It is then known as 'conditioned PFA', which can be stockpiled if it is not required immediately. At certain power stations PFA is transported hydraulically to lagoons from where it can be later reclaimed; it is then known as 'lagoon PFA'.

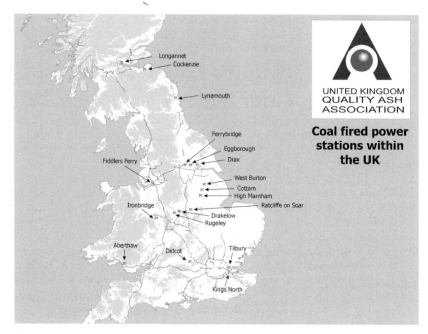

Fig. 36. Location of coal-fired power stations 1998 (map courtesy of the United Kingdom Quality Ash Association (UKQAA))

At some power stations FBA is washed out to lagoon storage with PFA. Due to separation during sedimentation in the lagoons, lagoon ash is more variable in composition and is very much coarser.

Composition of PFA

PFA consists of glassy spheres together with some crystalline matter and a varying amount of carbon. The overall colour ranges from almost cream to dark grey and is affected by the proportions of carbon, iron and moisture. The three predominant elements in PFA produced by burning British coals are silicon, aluminium and iron, the oxides of which together account for 75–95% of the material. Such ashes are known as alumino-silicate fly ash. There is also another variety, known as sulpho-calcitic fly ash, which is produced by the combustion of coal with a high limestone and sulphur content. Sulpho-calcitic ashes have a high lime (CaO) content; they can therefore have hydraulic properties because of the pozzolanic reaction between the components of the ash. A draft European Standard for sulpho-calcitic ash for use as a hydraulic binder specifies that the total calcium oxide content shall be between 37% and 58% with the free lime between 18% and 31%. Typical chemical analyses of British (alumino-silicate) ashes are given in Table 24.

Mineralogical analyses of these ashes show that the silicon is present partly in the crystalline form of quartz (SiO_2) and in association with

Table 24. Chemical composition and physical properties of PFA (UKQAA 1998b, Sherwood and Ryley 1966)

Chemical component	Average by weight (%)
SiO_2[a]	45–51
Al_2O_3	27–32
Fe_2O_3	7–11
CaO	1–5
MgO	1–4
K_2O	1–5
Na_2O	0.8–1.7
TiO_2	0.8–1.1
SO_3[b]	0.3–1.3
Cl	0.05–0.15
Loss on ignition	1.3–3.2
pH	9–12
Physical properties	
Specific surface	1900–5000 cm^2/g
Particle density	1.90–2.37 Mg/m^3
Bulk density	1.2–1.7 Mg/m^3

[a] Water-soluble sulphate
[b] The figures for SiO_2 do not necessarily refer to free silica but to silicon present as silicates of varying compositions

the aluminium as mullite ($3Al_2O_3.2SiO_2$), the rest being in a non-crystalline glassy phase. The iron appears partly as the oxides magnetite (Fe_3O_4) and haematite (Fe_2O_3), the rest is in a glassy phase. The greater proportion of PFA is in fact a glass with the glass content varying between 65% and 90%.

Physically, PFA is a fine powder which bears a close resemblance to Portland cement in general fineness and usually also in colour. The range of physical properties of some samples of hopper ash are given in Table 24. Figure 37 gives the particle size distributions of several samples of PFA ashes which shows that they are predominantly silt size. Although some of the power stations producing these ashes are no longer operational the results still give a good indication of the range of particle size to expect. For comparison the particle size distributions of two lagoon ashes are also included in Fig. 37, showing the generally much coarser gradings of these ashes.

PFA presents few problems of a chemical nature when used in road construction, except possibly for leachates which are considered in Part 3. It invariably contains sulphates, occasionally in high concentrations. However, because of the low permeability of compacted PFA the presence of sulphates has not been found to be a problem in practice.

Most fresh ashes are highly alkaline, due mainly to the presence of free lime. Carbonation reduces the alkalinity with time but the ash remains well on the alkaline side of neutrality. This does not cause any problems except that metals such as aluminium which corrode in alkaline environments should not be allowed to come into contact with PFA.

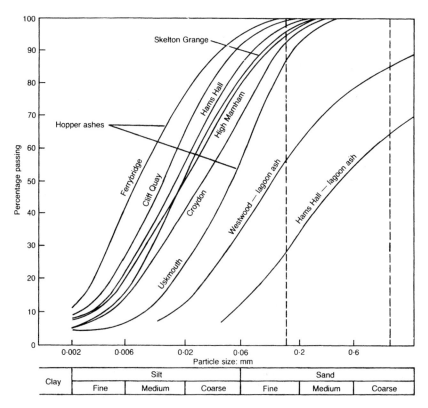

Fig. 37. Particle size distribution of samples of PFA (Sherwood and Ryley 1966)

Specifications for PFA

As PFA finds considerable uses in cement and concrete, specifications have been prepared giving the requirements for its use in such circumstances. BS 3892: Part 1 (1997) gives the specification of PFA for use with Portland cement. Other parts of BS 3892 deal with the use of PFA in grouts and precast concrete products which are outside the scope of this book. Part 1 will eventually be superseded by BS EN 450 which, like BS 3892, limits the loss on ignition to 7% but permits a greater range of fineness. The supplier nominates a target fineness between 0% and 40% retained on the 45 μm sieve and the PFA must not exceed ±10% of this target value.

Composition of FBA

FBA is the coarser agglomerated material recovered from the bottoms of the combustion chambers of power station boilers fired with pulverized fuel. In appearance it ranges from a highly vitrified, glossy and heavy material to a lightweight, open-textured and more friable type. The precise nature of the material depends on the boiler plant and coal type. It may occur, mixed with PFA, in lagoon ash.

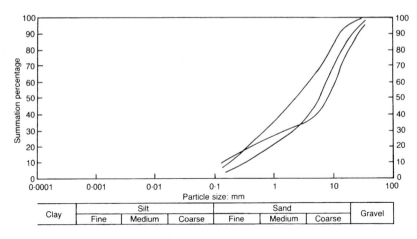

Fig. 38. Particle size distribution of three FBAs (CEGB 1972)

Chemically FBA is very similar to PFA, but in its physical properties it differs entirely, being a coarse granular material ranging in particle size from fine sand to coarse gravel (Fig. 38). The grading makes it potentially suitable as a selected granular fill but because the particles have a porous structure they are relatively weak compared to most granular materials used in road construction.

Uses of PFA in road construction
Bulk fill
There are three types of PFA readily available for use as a fill material:

- *conditioned ash* – PFA taken directly from the silos at a power station to which a controlled amount of water is added to give the optimum moisture content and to assist in handling and compaction on site
- *stockpiled ash* – previously conditioned PFA which has been stock-piled prior to use
- *lagoon ash* – PFA which has been slurried and pumped to storage lagoons. It is then allowed to settle and drain before delivery. It is coarser than conditioned ash (see Fig. 37).

PFA is a valuable bulk fill material but because of its unusual properties it is given a separate classification (2E) in Class 2 of general cohesive fills listed in the Specification for Highway Works. Class 2E is specified as reclaimed material from lagoon or stockpile containing not more than 20% of FBA. Certain precautions have to be followed in constructing embankments from PFA and in consequence the Specification requires that it shall not be placed within the dimension described in the contract below sub-formation or formation. This is because:

(*a*)   the grain shape and particle size of PFA make the upper layers of PFA difficult to compact

(*b*)   freshly placed PFA behaves in a similar manner to silt and, if not protected, may liquefy under wet conditions
(*c*)   capping and sub-base materials tend to be relatively permeable and a layer of general fill over PFA is considered desirable to add some protection.

The Specification also requires the use of a starter layer of Class 6D material below PFA used as bulk fill. This is partly to provide a firm working platform for the construction of the PFA fill. It also functions as a capillary cut-off to inhibit the upward movement of ground water into the PFA. PFA is a silt-sized material (Fig. 37) with a high capacity for capillary rise, and the possible upward movement of water into the PFA before compaction has occurred can give rise to problems unless the starter layer is included.

Another distinctive feature of the use of PFA as described in the Specification is that, unlike other general bulk fill, it is compacted to an end-product specification. Most of the materials considered in this book have particle densities in excess of $2.3 \, Mg/m^3$. Conversely, most ashes produced from coal of British origin have particle densities well below this value. This low particle density is reflected in the results of compaction tests carried out on PFA which when compacted has a low density compared with most other materials used for mass fill (Table 25). This lightweight property is advantageous when fill material is required on highly compressible soils, and PFA is often specified in these situations.

However, according to Clarke (1992) the particle density of ashes from some British power stations has been increasing in recent years. Clarke claims that PFAs from such stations can no longer be classified as lightweight fill since their particle densities are similar to that of some soils. British coals have not suddenly changed their composition and the claim, if true, is probably due to the increasing use of imported coal.

Many ashes possess self-cementing properties when they are compacted. The result of this hardening, if it occurs, is that settlement within PFA fill is less than with other materials. This makes it particularly

Table 25. Maximum dry densities and optimum moisture contents for various fill materials (CEGB 1972)

Type of material	Typical results of a BS compaction test	
	Maximum dry density (Mg/m^3)	Optimum moisture content (%)
Gravel	2.07	9
Sand	1.94	11
Sandy clay	1.84	14
Silty clay	1.67	21
Heavy clay	1.55	28
PFA	1.28	25

useful as a selected fill material behind bridge abutments, where settlement can be particularly troublesome.

PFA can present unusual problems of sampling. For PFA from lagoons the particle size is likely to vary in the lagoon and the size becomes finer with increasing distance from the outfall. To maintain reasonably consistent gradings the material should therefore be excavated in batches at roughly constant distance from the discharge pipes. PFA can also be a problem when conditioned ash is delivered from more than one power station, as the different ashes have widely different properties. For this reason the specification requires that, for each consignment of ash, a record should be kept of the type and source of the material and the name of the power station from which it was obtained.

As a capping material

Although it may have self-hardening properties, PFA used alone is not suitable as a capping layer material. However, as can be seen from Table 26 it may be readily stabilized with cement and the Specification for Highway Works lists conditioned PFA as being suitable for this purpose (Class 7G, which, when stabilized with cement, becomes Class 9C). The specification requires the PFA to be conditioned ash direct from the power station dust collection system and to which a controlled quantity of water has been added.

In recent years the United Kingdom Quality Ash Association (UKQAA) has introduced the concept of using lime-stabilized ash as a

Table 26. Compressive strengths of cement- and lime-stabilized PFA

Results of Sherwood and Ryley (1966)

Ash sample No.	Compressive strength (MPa)				
	With 10% cement		With 10% lime		
	7 days	28 days	7 days	28 days	56 days
1	8.35	13.8	1.58	9.95	ND
2	5.32	13.1	0.55	5.90	ND
3	4.35	8.37	1.03	3.37	ND
4	5.62	7.33	3.44	10.8	13.0
5	3.38	7.13	0.83	3.79	8.02

Results of UKQAA Technical Data Sheet 6.5 (UKQAA 1998b)

	7 days	28 days	35 days	28 days +7 days in H_2O	91 days
Ash + 2.5% CaO	1.5	ND	4.0	3.3	6.0
Ash + 5% CaO	1.8	ND	4.0	3.3	7.3
Ash + 7% cement	3.0	4.0	ND	ND	6.0
Ash + 9% cement	6.0	8.0	ND	ND	9.0

ND – Not determined

capping material and for sub-base. The product is termed LFA (Lime Fly Ash) the results of their tests and of earlier tests by Sherwood and Ryley are summarized in Table 26. The results show the advantage of cement over lime in the initial stages of hardening but as the pozzolanic reaction between lime and PFA proceeds the strengths developed are comparable and may even exceed those produced by cement.

There is little to choose between the price of cement and lime so that the use of lime in preference to cement offers little economic advantage. However, the slower setting of lime-stabilized material lessens the need to complete compaction within the limits prescribed for cement-stabilized materials and lime-stabilized PFA has self-healing properties. Compared with cement-stabilization the setting of lime-stabilized materials is more temperature-sensitive and below 5 °C the reaction virtually ceases.

As a sub-base material
On its own PFA has no use as a sub-base material but Table 26 shows that, when stabilized with cement, it can be made to meet the requirements of CBM1 cement-bound sub-base material. The results in Table 26 also show that lime can be used instead of cement to give comparable long-term strengths; the use of fly ash bound mixtures (FABMs) as sub-base materials is considered in more detail later.

As an additive to cement-bound base materials
Dunstan (1981) showed that what he called 'rolled concrete', which was in effect lean concrete (CBM3) with a high addition of PFA, had many advantages over alternative materials for use in dam construction. The properties that were particularly useful were low heat of hydration, ease of compaction, low permeability and good bonding between layers. With the exception of the heat of hydration, which does not create a problem in road construction, the improvement of the properties of lean concrete by the addition of PFA suggested that ash-modified lean concrete might have potential as an alternative to lean concrete (CBM3) for road base construction.

In view of this, the properties of ash-modified lean concrete as a road base material were evaluated (Franklin *et al.* 1982, Harding and Potter 1985, Potter *et al.* 1985). This work showed that the pozzolanic properties (see below for definition) of PFA allowed some of the cement in lean concrete to be replaced with PFA, and although this reduced the early strength the long-term strength was increased. However, the water-reducing properties of PFA could be used to reduce the effect on the early strengths and it was possible to achieve almost any desired strength requirement. This is illustrated by Fig. 39 in which the curve labelled LC is the strength–age relation for normal lean concrete made from a flint–gravel mixture. The curves labelled L1–L5 represent ash-modified lean concretes with the PFA/cement ratios and water/ash + cement ratio (on a volume basis) shown in the accompanying table.

The results in Fig. 39 show that the rate of development of strength of ash-modified lean concrete is quite different from that of the control. By

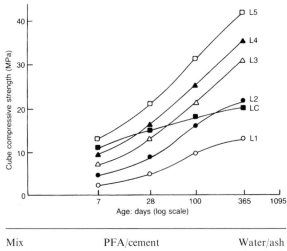

Mix	PFA/cement	Water/ash + cement
L1	4.00	1.62
L2	3.99	1.22
L3	2.76	1.32
L4	1.50	1.60
L5	1.50	1.35
LC	—	3.24

Fig. 39. Strength–age relations of PFA-modified lean concrete prepared from a flint-gravel aggregate (Franklin et al. 1982)

judicious selection of the mix proportions it is possible to design ash-modified lean concrete that has a strength comparable with that of normal lean concrete at the desired age, but at earlier ages the strength will tend to be lower and the long-term strength much higher. This means that, in principle, ash-modified lean concrete road bases could be designed to have a low early life strength, to produce fine, closely-spaced cracking with good aggregate interlock, and a high in-service strength to ensure good performance under traffic with less likelihood of cracks propagating through the surfacing.

The water-reducing properties of PFA and the spherical shape of its particles also influence the compactibility of ash-modified lean concrete. It is easier to compact than normal lean concrete and the reduction in moisture content results in higher compacted densities being obtained. This means that it can be laid in greater thicknesses than normal lean concrete and satisfactory compaction has been obtained with layers up to 300 mm thick (see Fig. 40).

As an additive to concrete
PFA has been recognized for many years as a valuable material for modifying and enhancing the properties of concrete. The major benefit to be

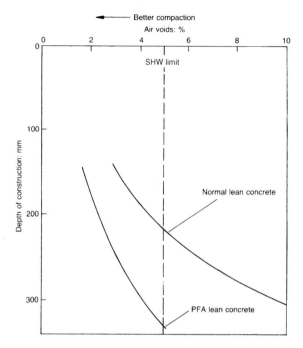

Fig. 40. Density gradients in pavement layers constructed with normal lean concrete and PFA-modified lean concrete (Potter et al. 1985)

gained results from the pozzolanic properties of PFA, that is, its reaction with calcium hydroxide produced as one of the hydration products of Portland cement. As this reaction is a consequence of the hydration of cement, it can contribute to the strength of concrete only after hydration has occurred. When ash is used as a partial replacement to cement, the initial strength of the concrete is therefore reduced but the final strength level often exceeds that of a conventional, unmodified mix.

Apart from its pozzolanic action, the addition of PFA to concrete also has significant physical effects. The effects arise partly because PFA contains a large proportion of spherical particles that give a plasticizing action that affects the water demand. In addition, because of the lower particle density of PFA (about $2.4\,g/cm^3$ compared with $3.15\,g/cm^3$ for cement) the introduction of PFA alters the volume of fine material (and hence the volume of cementitious material) by about 10%.

The combined effect of particle shape, grading and particle density causes a substantial reduction in the water demand for concretes containing PFA. In addition to this change, the increased volume of cementitious powder with better packing creates a structure which has the potential for much higher long-term strength, lower permeability and increased resistance to chemical attack. A summary of the effect of PFA on the properties of concrete is given in Table 27.

Table 27. Summary of the effect of PFA on the properties of hardened concrete (Bamforth 1992)

Property	Effect of PFA
Fresh concrete	
Water demand	Reduced with increasing proportions of PFA
Workability	At a constant water/cement ratio, the slump of PFA concrete is increased. At a constant slump the true workability of PFA concrete, e.g. response to vibration is increased
Placeability	Based on a qualitative assessment of performance on various construction projects, the addition of PFA improves placeability
Bleed and settlement	Limited test data show reduced bleed. No specific reports from site of significant differences between ordinary Portland cement (OPC) and PFA mixes; however, high flow mixes have been used with no bleed problems
Setting time	Increased by the use of PFA, typically by 1 hour at 30% level, 2 hours at 50%
Finishing	Can be delayed due to extended setting time
Hardening concrete	
Standard cure strength	Early strength is reduced but long-term strength is increased
Heat-cycled strength	For equivalent grade mixes, PFA concrete will have a higher heat-cycled strength at 28 days
Air-cured strength	Air-curing affects OPC and PFA concretes to the same extent
Load deformation	Reduced by about 20% at 30% PFA. Creep affected to a greater extent than elastic deformation
Drying shrinkage	Reduced
Durability	
Sulphate resistance	Increased compared to OPC concrete
Acid resistance	Likely to be increased due to lower permeability
Water & gas permeability	Comparable with OPC concrete at 28 days but reduced to a greater extent with age
Carbonation	Unaffected in structures, although short-term laboratory results sometimes indicate an increase
Chloride ingress and salt attack	Chloride ingress is reduced as is deterioration under cyclic salt exposure
Electrical resistivity	Increased substantially

In Portland–pozzolan cements

A pozzolana may be defined (ASTM C595-76 1976) as 'a siliceous, or siliceous and aluminous material, which in itself possesses little or no cementitious value but will, in finely divided form and in the presence of moisture, chemically react with calcium hydroxide at ordinary temperatures to form compounds possessing cementitious properties'. Many naturally-occurring and artificial materials have pozzolanic properties. The natural pozzolanas are for the most part materials of volcanic origin. The artificial pozzolanas are mainly products obtained by the heat treatment of natural materials such as clays, shales and certain siliceous rocks, and PFA.

Calcium hydroxide is one of the hydration reaction products of Portland cement. It contributes nothing to the strength of cemented products and can be a potential source of instability. The addition of a pozzolana to ordinary Portland cement can therefore be beneficial by reacting with the calcium hydroxide released by the hydrating cement to produce further cementitious material. Many of the pozzolanas mentioned above may be employed in the manufacture of Portland-pozzolan cements (ASTM C595-76 1976) but PFA is the most important and specifications exist for both the properties of PFA used in cement, in concrete and for Portland PFA cement.

As alternatives to Portland cement in cement-bound sub-bases and bases
Mixtures of lime and PFA may be used for stabilizing non-cohesive granular materials and, with the most reactive ashes, the lime/PFA mixtures compare favourably in price and performance with Portland cement (Sherwood and Ryley 1966, Gaspar 1976). However, the presence of clay in lime/PFA stabilized soils has a detrimental effect. This is probably because the lime reacts preferentially with the clay fraction to produce products with weaker cementitious properties than the lime–PFA reaction products.

This early work has been greatly extended in recent years to give a family of Fly Ash (PFA) Bound Mixtures which go under the acronym of FABMs. Examples of these are given in Table 28.

Lime fly ash bound granular material (GFA) has the most potential for use at sub-base level. It is a slow hardening mixture (see Fig. 41) which progresses from an unbound granular material into a bound pavement layer. It is claimed that this has the following advantages.

- In the short-term GFA has a handling time of many hours and thus increases the flexibility of unbound granular paving materials.
- In the long-term, and depending on the aggregate, GFA develops significant elastic stiffness (10–30 GPa) and tensile strength (1–3 MPa) to produce a pavement material (Fig. 42) with the performance and durability of bituminous- and cement-bound materials.
- The slow reaction rate realizes extended workability, permits immediate accessibility to site traffic and the capacity of self-healing.

Against this is the fact that the ultimate structural characteristics may take a long time to be achieved and the slow reaction time limits its use in the UK to the period April–October so as to ensure that adequate frost-resistance is achieved before the onset of winter.

Most PFAs contain free lime in the form of CaO. In the presence of water this can react with the siliceous components of the ash so that most ashes have at least some self-hardening properties. Sulpho-calcitic ashes contain sufficient free lime for them to be used as cementitious materials on their own. In southern France an ash is produced which contains up to 25% free lime. Gravel stabilized with 4% of this ash has been widely used for road construction in the area of the power station (Gaspar 1976). Similar ashes occur in Poland and are used to stabilize gravel for sub-base and base construction.

Table 28. Examples of fly ash bound mixtures (FABMs) (UKQAA 1998b)

Type of FABM	Abbreviation	Conditioned PFA (%)	Lime (%)	OPC (%)	Graded crushed coarse material (%)	Sand (%)	Soil (%)	Other material (%)	Typical water content (%)	Normal test age (days)
Lime PFA	LFA	93–97	3–7						15–25	90
Lime–gypsum PFA	LFA	91	4					5% gypsum	15–25	90
Cement–PFA	CFA	90–95		5–10						28
Lime–PFA granular material	GFA	8.5–13	1.5–3		50–55	40–45			6–8	90
Lime–PFA granular material	GFA		1–1.5		50–55	40–45		4–6% dry PFA	6–8	90
Cement–PFA granular material	GFA	3–6		1–3	50–55	40–45			6–8	28
Slag–PFA granular material	GFA	5–7	0–2		50–55	30–40		5–7% GBS[a]	6–8	90
Lime–PFA sand	SFA	9–12	2–4			84–89			Approx. 10	90
Cement–PFA sand	SFA	6–8		2–4		88–92			Approx. 10	28
Lime–PFA soil	EFA		1–2				90–93	6–8% dry PFA	Depends on soil	90
Cement–PFA soil	EFA	3–6		2–4			91–94		Depends on soil	28

[a] GBS – granulated blast-furnace slag

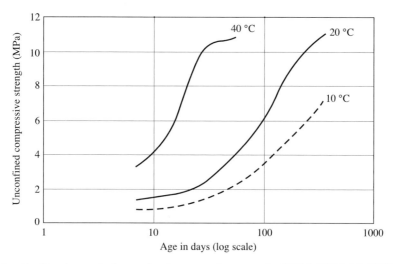

Fig. 41. Development of unconfined compressive strength of GFA (UKQAA 1998b)

Uses of FBA in road construction

FBA is a coarse granular material ranging in particle size from fine sand to coarse gravel (Fig. 38). The grading makes it potentially suitable as a selected granular fill and as a granular sub-base material, but because

Fig. 42. Placement of GFA base material on A52, Froghall, Staffs (photograph courtesy of UKQAA)

the particles have a porous structure they are relatively weak compared to most granular materials used in road construction. This means that in those instances where a requirement for particle strength in terms of a minimum TFV is specified (see Fig. 7) it will generally be unable to meet the necessary strength requirements. Dawson and Bullen (1991) found that all of the samples of the FBA studied by them had TFVs below 50. However, compaction trials showed that particle degradation of FBA was small and barely changed the amount of fine material in the whole. They therefore concluded that the TFV falsely condemned FBA on the basis of its degradability. This suggests that for minor road-works the TFV requirement could be abandoned if experience showed that the stability of the material after compaction was satisfactory for the purpose being considered.

9. Rubber

Occurrence

In the UK about 37 million car and lorry tyres (380 000 tonnes) reach the end of their lives every year (Environment Agency 1998). In 1996 of these it was estimated that about:

- 30% were re-treaded for re-use on road vehicles
- 27% were used for energy recovery
- 30% were disposed to landfill
- 3% were physically re-used for other purposes, e.g. crash barriers.

It is anticipated that the number of tyres for disposal will increase whilst at the same time the option of disposing them to landfill will be drastically reduced. This means that there will be an increasing need to find alternative uses for the tyres. Tyres have a high energy content and there is scope to increase the number used as fuel. Even so there is serious concern about the number of tyres for which it will be difficult to find an outlet and which, if illegally dumped, could create a fire hazard and have a high potential for pollution.

Uses in road construction

Less than half the weight of a scrapped tyre consists of rubber so that the figures given above exaggerate the amount of waste rubber that is available. Even so the amounts available are not insignificant and scrap tyres can be turned into crumb rubber which has some potential uses in road construction. The production process involves crushing and grinding the tyres using either conventional methods or a cryogenic crushing process (where the tyres are put into a liquid nitrogen bath before crushing), the separation of casing and fibres, and then crushing again to obtain the desired grain size.

In bituminous mixtures

Experiments to improve the properties of bitumen by the addition of rubber have been in progress for nearly 100 years. These have mostly been carried out with natural rubber and have shown that the addition of a small proportion of rubber to a bitumen binder has a strong influence on the rheological properties in a manner that is likely to result in improved performance. Full-scale trials have shown that when rubberized binders are used in road surfacings, the elasticity of the binder is increased so that such materials when used in road surfacings show a greater resistance to cracking and deformation under traffic loads (Thompson and Szatkowski 1971). However, despite the proven advantages the amount of natural rubber used in roads is very small and the potential for the use of waste rubber must be seen in this light. Moreover, the higher proportion of non-rubber constituents in waste rubber (carbon black, etc.) means that higher amounts have to be used – this can cause problems of blending; in addition waste rubber is variable in composition.

Most of the work on using waste rubber in bituminous pavements has been done in the USA where the following two types of mixtures containing rubber are recognized.

Asphalt Rubber Concrete (ARC) is a hot-mix asphalt concrete in which ground rubber from tyres is mixed with asphalt cement for use as a binder for the mineral particles in the mix. It is claimed that it is superior to conventional mixtures with respect to crack resistance.

Rubber Modified Asphalt Concrete (RUMAC) uses ground rubber to replace 2–5% of the mineral aggregate in hot-mix asphalt concrete.

In French drains

Filter (French) drains are often installed at the edge of the road in order to facilitate rapid removal of water from the road surface and to prevent water flowing onto the road. Research by Carswell and Jenkins (1996) showed that recycled rubber could be used in place of the aggregates normally used in French drains. This work showed that:

- acceptable designs of binder-bound rubberized materials can be made which have cohesive strength
- various sources of scrap tyres may be used provided a suitable grading can be achieved
- permeability tests showed that rubber in French drains performs similarly to conventional granular materials, with voids allowing surface water to pass rapidly through
- it can be permanently coloured to blend with its surroundings
- it is environmentally friendly and can be recycled
- it cures the problem of loose stones on sliproads.

10. Slags

Blast-furnace slag

Occurrence

Blast-furnace slag is a by-product obtained in the manufacture of pig-iron in a blast furnace, and is formed by the combination of the earthy constituents of the iron ore with the limestone flux. Iron ore is a mixture of oxides of iron, silica and alumina. The chemical reactions in the blast furnace reduce the iron oxides to iron while the silica and alumina compounds combine with the calcium of the fluxing stone (limestone and dolomite) to form the slag. The chemical reactions occur at temperatures between 1300 °C and 1600 °C produced by the burning of coke which is fed into the furnace along with the ore and the limestone and dolomite. When preheated air is blown into the furnace the oxygen combines with the carbon of the coke to produce heat and carbon monoxide. The iron ore is reduced to iron mainly through the reaction with the iron oxide to produce carbon dioxide and metallic iron. The limestone and dolomite flux is calcined by the heat and dissociates into calcium and magnesium oxides and carbon dioxide. The oxides of calcium and magnesium combine with the silica and alumina of the iron ore to form slag. Thus compounds of lime–silica–alumina and magnesia are formed which collect in a molten stratum above the molten iron in the furnace.

The physical appearance and mineral structure of blast-furnace slag depends largely on the method by which it is cooled. About 70% is air-cooled, which produces a rock-like material with a crystalline structure which when crushed can be used as an excellent substitute for natural aggregates. About 25% is subjected to a water-cooling process where the rapid solidification produces a granulated glassy material. Two other forms exist: foamed slag produced by generating steam within the molten slag from jets of water arranged around the bottom and sides of the pond into which the slag is poured, and pelletized slag, which is similar to granulated slag.

The proportion of slag-to-iron produced by the blast furnace varies with the richness in iron of the ore. In the UK the amount of indigenous ore used for iron production has steadily fallen; by 1987 it had fallen to negligible proportions and all the ore used was imported. The imported ore has a much higher iron content than home-produced ores and this means that even if the production of iron had remained static the amount of slag produced would have fallen. In fact, iron production

Table 29. Chemical composition of blast-furnace slag (Lee 1974)

Component	Range (% by mass)
CaO	36–43
SiO_2	28–36
Al_2O_3	12–22
MgO	4–11
Total sulphur (as S)	1–2
Total iron ($FeO + Fe_2O_3$)	0.3–2.7

has also fallen, so less than 4 million tonnes of blast-furnace slag was produced in 1996, which is about half that produced ten years earlier. Moreover it is continuing to fall as further closures of steel works have recently (2001) been announced. This means that the UK is no longer a major producer of slag.

Composition
The crystalline materials that form blast-furnace slag are compounds of the oxides of calcium and magnesium with silica and alumina. Sulphur is also present as sulphides and sulphates, with minor amounts of iron. The range in the chemical composition of slag is given in Table 29.

Specifications for blast-furnace slag
For most purposes blast-furnace slag is generally regarded as being at least as good as natural aggregates. But, as it differs from these in its chemical and physical properties, the specification and testing methods used for natural materials do not necessarily apply.

BS 812 (1988) is the principal British Standard relating to the testing of aggregates in an unbound form. The physical tests in the standard can, for the most part, be used not only for naturally-occurring aggregates but also for related materials such as slags and crushed concrete. However, the chemical composition of manufactured materials, such as blast-furnace slag, may be markedly different from natural materials, so the chemical tests in BS 812 may not be applicable. BS 812 is due to be superseded by comparable European Standards by the end of 2003.

In the case of blast-furnace slag the chemical requirements for its use in construction are given in BS 1047: 1983 'Air-cooled blast-furnace slag for use in construction'. This gives the requirements for the physical and chemical properties of the slag. In the case of the physical properties this is done by reference to the appropriate tests in BS 812. In the case of the chemical properties the standard gives requirements and methods of determining stability (soundness), total sulphur content and water-soluble sulphate content. BS 1047 will also eventually be superseded by European Standards and BS EN 1744-1: 1998 'Tests for chemical properties of aggregates' contains methods for testing blast-furnace slag that are very similar to those included in BS 1047.

Soundness. Two forms of instability can occur. Iron unsoundness is very rare and there is a suspicion that the requirements given in BS 1047 and BS EN 1744-1 are included because it is so easily detected. It arises when partially-reduced iron oxides in the slag oxidize. This is an expansive reaction which causes the slag to disintegrate. It is detected by immersing twelve pieces of slag in water for a period of 14 days and observing whether any of the particles crack or disintegrate.

Calcium disilicate unsoundness, also known as 'falling' and misleadingly as 'lime unsoundness', arises from a phase change when the meta-stable beta form of calcium disilicate changes to the gamma form. The phase change is accompanied by a 10% increase in volume. Calcium disilicate cannot form in significant amounts when the ratio of CaO and MgO to the SiO_2 and Al_2O_3 and S is kept within the limits given in BS 1047. Two conditions given are:

$$CaO + 0.8MgO < 1.2SiO_2 + 0.4Al_2O_3 + 1.75S \text{ (\% by mass)}$$
$$CaO < 0.9SiO_2 + 0.6Al_2O_3 + 1.75S \text{ (\% by mass)}$$

Slags which satisfy either one or both requirements are regarded as sound. Slags which fail to meet these conditions are not necessarily unsound and if the analytical test fails to give a positive answer the decision on whether or not they are unsound is then made solely on the basis of microscopic examination.

BS EN 1744-1 also includes a method for detecting unsoundness but instead of relying on chemical analysis it specifies examination of particles of slag under ultra-violet light – the aspect and colour of fluorescence enable the detection of slags suspect with respect to silicate disintegration.

Sulphates. Slags used for concrete aggregate have to satisfy the requirements for stability and are also required to have a total sulphur content of not more than 2% (as S) and a sulphate content of not more than 0–7% (as SO_3). Slags used in an unbound form have to meet the stability requirements and to have a soluble sulphate content of not more than 2 g/litre.

Other problems. Before the first edition of BS 1047 in 1942 blast-furnace slag had a poor reputation. This problem was effectively dealt with by the new standard so that virtually all new slag production could be adjusted to comply with the standard. The introduction of BS 1047 allowed new blast-furnace slags to be used with confidence and in 50 years of utilization there has been no evidence to suggest that this confidence has been misplaced. Moreover, problems with the stability and sulphate content of current blast-furnace slag are rare.

Apart from concern about sulphates and stability, the scope of BS 1047 (1983) states that it 'excludes requirements for blast-furnace slags from haematite pig-iron (except when dolomite is used as the flux in its manufacture)'. This exclusion has been in the standard for a long time and it is now unclear why it was first introduced. It is in any case highly ambiguous because it can be read in any one of four ways, as follows.

(a)   The slags referred to could not possibly meet the requirements of BS 1047 so that it is a waste of time to test them.

(b)   Whether or not such slags meet the requirements of BS 1047 they are inherently unsuitable and should not be used.

(c)   Such slags may well be satisfactory for use but the tests in BS 1047 cannot be applied to them.

(d)   The slags are known to be satisfactory and do not need to be tested.

In addition to the ambiguous way in which the clause is worded there is a further ambiguity in what is actually meant by 'slags from haematite pig-iron': how can the ore from which the slag was produced be identified? The difference between haematite and other slags, as seen by mineralogy and chemical analysis, is likely to be small compared with the variation within each group. It therefore seems probable that (a) was the reason for the exclusion and if the slag meets the requirements of BS 1047 for stability no problems are likely to arise from this cause provided that it is blast-furnace slag from current production.

However, old deposits of slag are known to give problems which are summarized in BS 6543: 1985 'Guide to the use of industrial by-products and waste materials in building and civil engineering'. This states that 'slags from old blast-furnace slag banks are not recommended as fill under buildings unless thoroughly sampled and tested, as they may contain slag wastes other than blast-furnace slag or other industrial wastes. Moreover, there is evidence that some old, partially vitrified slags may react and expand if exposed to sulphate solutions originating either in the ground water or other components of the fill'.

The presence of slags other than blast-furnace slag within a deposit would clearly mean that the requirements of BS 1047 (which is specific to blast-furnace slag) would not necessarily apply. This partly explains the warning about old slags given in BS 6543, but the standard is not clear as to whether or not a blast-furnace slag which complied with BS 1047 would suffer from the problems associated with sulphates.

Collins (1993) investigated a case of expansion in an old deposit of blast-furnace slag which was the subject of internal sulphate attack. This involved the formation of the mineral thaumasite $[Ca_3Si(OH)_6]_2$-$SO_4(CO_3)_2 24H_2O]$; this mineral is similar to, and isomorphous with, ettringite, the mineral which forms in the sulphate attack of concrete. In this instance the slag did not comply with BS 1047 and Collins concluded that compliance with this standard was a sufficient guarantee of stability. There have been other reported cases of expansion of blast-furnace slag attributable to sulphate reactions but there is no published evidence of whether or not these slags would have been rejected by BS 1047.

Uses of blast-furnace slag in road construction

Air-cooled slag.   Some of the physical properties of air-cooled blast-furnace slag are summarized in Table 30. When the molten slag from the blast furnace solidifies it can be crushed and screened by normal

Table 30. Physical properties of air-cooled blast-furnace slag (after Lee 1974)

Particle density	2.38–2.76 Mg/m^3
Bulk density	1150–1440 kg/m^3
Water absorption	1.5–5% (by mass)
TFV	70–160 kN
Aggregate impact value	21–42%
Aggregate abrasion value	5–31

quarrying procedures to produce a material which satisfies the grading requirements of most specifications. When crushed and screened, the physical properties of the slag make it particularly suitable as an aggregate, both coated and uncoated. It breaks to give a consistently good cubic shape and it has a rough surface texture giving good frictional properties and good adhesion to bituminous and cement binders. It is therefore widely used in civil engineering construction as a substitute for naturally-occurring aggregates. This is reflected in the Specification for Highway Works which mentions blast-furnace slag by name as being suitable for use in all levels of the road pavement structure, provided that it meets the relevant requirements.

It may be used in an unbound form as selected granular fill and as a granular capping and sub-base material. The Specification allows for blast-furnace slag to be used as a granular fill for all purposes except where it is used under water or comes into close proximity to metals or as a starting layer below PFA. The use of slag below water could give rise to problems with water pollution and there is a risk of corrosion if slag comes into contact with metals that corrode in an alkaline environment. The exclusion of slag for use as a starter layer below PFA is presumably because of doubts over its drainage characteristics.

There are even fewer restrictions on the use of blast-furnace slag as an unbound granular capping and/or sub-base material, and the self-cementing action of the slag means that, in many ways, it is superior to its naturally-occurring counterparts. Although it can be readily stabilized with cement to meet the requirements of CBM1 and CBM2 sub-base materials, it is seldom necessary to do so.

Blast-furnace slag is permitted for use, with few restrictions, as a CBM3 and CBM4 road base material and as a concrete aggregate. It is also used as an aggregate in bituminous mixtures where its rough texture and relatively high porosity, together with its alkaline activity, produces good adhesion characteristics particularly in the presence of water.

Blast-furnace cements. Table 29 shows that in basic chemical composition slag contains the same elements as Portland cement. It is not a pozzolana nor is itself cementitious, but it possesses latent hydraulic properties which can be developed by the addition of an activator such as lime or another alkaline material.

Table 31. Energy content of hydraulic binders (OECD 1984)

Material	Energy content as produced (MJ/t)
Ordinary Portland cement (97% clinker)	5000
Granulated slag	40
PFA	25
Cement with high slag content (15% clinker)	1900
Cement with low slag content (65% clinker)	3800

Many countries use ground granulated blast-furnace slag (GGBFS) in the manufacture of so-called blast-furnace cement. The percentage of slag in the Portland cement clinker, with which it is ground, may be very low or as high as 85%. The addition of materials such as blast-furnace slag or PFA to Portland cement, even in high proportions, does not impair its hydraulic properties. There may be a decrease in the early strength of concretes made from such cements but the long-term strength is not affected and the incorporation of slag or PFA can give rise to considerable savings in the energy consumed in the production of hydraulic binders (Table 31).

The energy requirements for the production of Portland cement are very high in relation to those for other components of concrete and cement-bound materials. It follows, therefore, that the use of slag and pozzolana cements can lead to significant savings in energy. A survey of the consumption of energy in road building by the OECD (1984) showed that a cement-bound gravel stabilized with 3.5% of Portland cement and with 3.5% of a high-slag cement consumed 268 MJ/t in the first case and 157 MJ/t in the second.

As an alternative to the addition of blast-furnace slag at the works, GGBFS is also available for addition to concrete made on the site. GGBFS is cheaper than Portland cement, so economies can be made.

Lime–blast-furnace slag mixtures.   As mentioned above, blast-furnace slag may be added to Portland cement clinker to manufacture Portland blast-furnace cement. This cement may be used for stabilization in the same proportions as ordinary Portland cement, and the properties of the stabilized material are much the same whether ordinary Portland cement or blast-furnace cement is used. Apart from its use in this manner, the cementitious properties of blast-furnace slags have been developed in France under the title of '*grave-laitier*' (gravel-slag) to stabilize gravel and sands for sub-base and base construction.

Granulated slag does not possess any hydraulic properties until it is activated. In the grave-laitier process this activation is achieved by small additions of hydrated lime. The reactivity of the slag varies between works and is measured by a so-called alpha (α) coefficient which is defined as

$$\alpha = S \times P \times 10^{-3}$$

where S is the Blaine specific surface (cm^2/g) of the natural fines of the granulated slag ($<80\,\mu$m) and P is the friability property determined by mixing in a ballmill 500 g of granulated slag with 1950 g of porcelain beads (diameter 18–20 mm) and subjecting the mix to 2000 revolutions at a rate of 50 rev/min. The factor P, defined as a percentage finer than 80 µm, is determined after sieving and washing.

On the basis of this coefficient, granulated slags are divided into four classes:

- Class 1: $\alpha < 20$, not used in road construction
- Class 2: $\alpha = 20$–40, the most frequently used
- Class 3: $\alpha = 40$–60, reserved for materials that are difficult to handle
- Class 4: $\alpha > 60$, used only exceptionally.

In France, grave-laitier is produced in mixing plants: it consists of a mixture of gravel with 15–20% of granulated slag together with 1% of hydrated lime (as the activator) and a moisture content of 10%. The strengths obtained from grave-laitier are only half those that would be obtained if Portland cement were used. However, granulated slag is cheaper than cement and grave-laitier has considerable advantages, which are summarized (OECD 1977) below, over cement-stabilized gravel as a construction material.

(a)   A relatively large quantity of granulated slag facilitates a homogeneous distribution of the binder in the mass. Part of the slag remains available, enabling a renewed setting (self-healing) should cracking occur.

(b)   Grave-laitier takes a relatively long time to set, allowing several days of storage without difficulty. It also allows flexible organization of the roadworks, each machine operating individually and at its maximum output.

(c)   Roadworks equipment can be allowed to circulate over the grave-laitier as soon as it is laid. Post-compaction due to traffic is good. The material is suitable for strengthening purposes while traffic is maintained.

(d)   In the case of heavy rain, excess water is simply allowed to drain off before proceeding with compaction. If necessary, materials may be respread, allowed to dry out and then recompacted.

(e)   The setting process is halted under frost, but setting recurs once normal temperatures are reached.

(f)   Strength takes a long time to build up fully (one year or longer) and is not affected by an initial delay in setting.

(g)   The slow rate of setting allows the moduli of the grave-laitier layers to increase progressively with the consolidation of the subgrade and increasing traffic.

Grave-laitier is the most widely used road base material in France and it is estimated that 65% of French roads have a pavement layer composed of grave-laitier. An extension of the process is to use air-cooled crushed slag as the coarse component of the mixture when, to use French terminology, the 'grave' is replaced by slag and the product is known as 'grave-laitier tout laitier'. A useful summary of French procedures on the use of hydraulic binders in road construction is available in both French and English (Ministere des Transports 1980).

A technique similar to the 'grave-laitier' process is used in South Africa where GGBFS is known, somewhat confusingly, as milled granulated blast-furnace slag (MGBS). South African specifications (NITRR 1986) give a ratio of four parts of MGBS to one of hydrated lime as the optimum proportions, but suggest that equal parts of MGBS and lime are often used since this is a convenient ratio in practice even though more lime may be used than is needed.

Instead of using air-cooled blast-furnace slag as the coarse component, phosphoric slag may be used. Phosphoric slag is the by-product of phosphorus manufacture but it has a similar chemical composition to blast-furnace slag. A blend of 85% phosphoric slag and 15% granulated blast-furnace slag is used as a road base material in Holland and is imported from there into south-east England where it has been similarly successful (Kent County Council 1985). When slag is used as the coarse component of the blend the hydraulic properties of the granulated slag do not require activation by the addition of basic activators such as lime.

Steel slag
Occurrence
The manufacture of steel involves the removal from the iron of excess quantities of carbon and silicon by oxidation and the addition of small quantities of other constituents that are necessary for imparting special properties to the steel. There are three main types of steel-making furnace that can be used: open hearth, basic oxygen (BOS) and electric arc (EAF), but only BOS and EAF slags are now produced in the UK.

Problems with using steel slags
The chemical composition of steel slag depends on the method of production (Table 32) but all, when cooled, give a product that resembles igneous rock and which would seem at first sight to be an excellent man-made aggregate. However, unlike blast-furnace slag, which is widely accepted

Table 32. Chemical composition of steel slags (Coventry et al. 1999)

	Chemical composition (% by weight)		
	BOS	Open hearth	EAF
CaO	42–47	33–51	31–50
SiO$_2$	13–16	9–19	11–24
Fe$_2$O$_3$	18–20	24–45	5–30
Al$_2$O$_3$	1–3	0.5–3	5–18
MgO	7–9	0.5–4	2–8
MnO	3–6	3–10	6–22

as an excellent substitute for natural aggregates, all types of steel slag are viewed with suspicion. This is because, when freshly produced, they can contain both free calcium oxide (CaO) and free magnesium oxide (MgO) – these, in the presence of water, hydrate to produce their respective hydroxides. Both hydration reactions are accompanied by an increase in volume but the reaction times can be considerable. Calcium oxide reacts violently with water so most of this is soon removed by hydration, however, some may be trapped inside the particles of slag or it may be in an unreactive form (hard-burnt) and in either case will therefore hydrate much more slowly. Magnesium oxide and hard-burned lime may take decades or perhaps even centuries to hydrate completely (Collins and Sherwood 1995).

EAF slags generally have a lower free oxide content than BOS slags but there is still a risk of expansion. The risk is much reduced by allowing the slag to stand in the open (weathering) but even when it has been weathered for a year problems have been known to occur in its use. In Great Britain only 10% of the BOS slag produced finds any use but about half of the EAF slag is used mainly as a bitumen-bound road aggregate in the upper layers of the road where any expansion that might occur can be tolerated and where the bitumen coating reduces the possibility of hydration. Steel slags are not recommended for use in road foundations and their use is actively discouraged in building construction. For example, a recent report (Collins and Sherwood 1995) sponsored by the Department of the Environment and which, ironically, has the aim of encouraging the use of waste materials says of all types of steel slag that 'Weathering does not remove sufficient of the free oxides to completely eliminate expansion, so these materials must, in their present form, be regarded as completely unsuitable for any use in construction, especially as underfloor fill and hardcore'. Later, when specifically discussing the use of hardcore, the report states that 'Steel slags are strongly discouraged. Every year millions of pounds of damage to buildings are caused by the misuse of steel slag'. A similar note of caution is contained in the British Standard relating to the use of waste materials (BS 6543: 1985).

Apart from the warnings given in these publications there is much published evidence that the use of steel slags in foundations can give rise to serious damage to buildings. Most of this originates from the 1970s which may suggest that the problem has lessened since then. This could be due to the greater amount of EAF slag now produced which is known to contain less free oxides and is therefore less of a problem than BOS slag. On the other hand it could indicate that, following these reports, people are now more cautious with the use of steel slags.

At the present time there are no UK specifications relating to the use of steel slag. Verhasselt and Choquet (1989) gave the following specifications for Belgian steel slags:

(a) Steel slag should not have a free lime (CaO) content of more than 4.5% at the time of production.
(b) Before use the slag should be allowed to weather for one year.

(c)   The maximum particle size should not exceed 20–25 mm.
(d)   Before use the volumetric stability should be checked by a volumetric swelling test.

Overcoming the problems
Not all steel slags are unsuitable for use and in all respects other than their potential chemical instability they make a good alternative to natural aggregates. Weathering reduces the problem of expansion but does not necessarily eliminate it. Determination of the total calcium and magnesium content of the slag is of little value because this gives no indication of the amount of free CaO and MgO that may be present. The general method of expressing the elemental analysis of aggregates, soils and related materials in terms of their oxides can be misleading because, for the most part, these elements will not be present as their free oxides although a low figure for total magnesium oxide would indicate that the free MgO content would be correspondingly low. The determination of the free CaO content of steel slag is not too difficult to carry out but free MgO is not so readily determined. In any case chemical methods are only an indirect route to finding whether or not the slag is likely to expand on contact with water over a period of time. Expansion tests are therefore preferable but the problem here is that the expansive reaction is slow and expansion tests carried our at 20 °C show that the slag may still be expanding after one year (Emery 1976). Emery recommended the use of elevated temperatures to accelerate the reaction and devised an expansion test carried out at 82 °C in which the expansion after 7 days was equivalent to the expansion after one year when the test was carried out at 20 °C. A German expansion test for steel slags involves testing the material at 100 °C and this method has been adopted as the basis of a European Standard test for steel slag which will be used throughout the European Union. The standard contains a let-out clause to the effect that 'Tests for the volume stability of steel slag aggregate, with satisfactory performance records are not necessary'. It is suspected that, in practice, producers will have difficulty in convincing users that their slags actually have a satisfactory performance record.

In parallel with the development of the expansion test the European Committee for Standardization (CEN) is preparing a European standard giving the requirements for 'Aggregates for unbound and hydraulic bound materials for use in civil engineering work and road construction'. According to this Standard aggregates are defined as any 'granular material used in construction, they may be natural, artificial or recycled', and it therefore embraces the use of steel slag. The Standard gives the requirements for the volume stability of steel slag when subjected to the expansion test. Slags with total magnesium oxide content of 5.0% or less are required to be tested for 24 hours and slags with more than 5.0% total MgO have to be tested for 168 hours (one week). The degree of expansion that can be accepted depends on the use of the slag, the most stringent requirement sets a maximum expansion by volume of 10% and the least sets a maximum expansion by volume of 5%.

The European Standards are in an advanced state of preparation; when published they will become mandatory throughout the European Union and will replace national standards relating to natural aggregates, slags and similar materials. As far as Britain is concerned the publication of the requirements for steel slags is an advance because at present their use in building construction is strongly discouraged. If a particular source of steel slag could be shown to meet the recommended requirements there could be no reason to reject it. However, this advance comes with a serious penalty because the testing regime for a material as variable as steel slag would have to be stringent. Not only does the composition of fresh slag from a steel works vary over time but the material in a stockpile will also vary depending upon how long it has been weathering in the stockpile.

Uses of steel slag in road construction
Although in many respects steel slags appear to be good-quality roadmaking aggregates, the possible presence of free lime (CaO) and free magnesia (MgO), with the consequent risk of expansion if and when these hydrate, severely limits their use in road construction and virtually excludes them from use as fill under structures.

Due to the possibility of long-term expansion steel slags are used, if at all, only in those situations where expansion is unlikely, as in the case of dense bitumen macadam, or where any expansion that does occur is not likely to be a serious problem, as in the case of surface dressing. Their main use, therefore, is in the upper bituminous layers of the road structure where the fact that the material is impermeable and the aggregate particles are coated with bitumen means that water would have difficulty in penetrating into the particles to cause any hydration. If any expansion did occur it would be limited to the upper layers which would cause less serious disruption than expansion occurring in the lower layer.

Non-ferrous slags
Occurrence
Apart from blast-furnace slag and steel slag, which have already been considered, small quantities of slag are produced from the smelting of other metal ores which may find an outlet in roadmaking in the vicinity of the works. In 1991 the amounts of such slags available were estimated (Whitbread *et al.* 1991) to be:

- Copper slag: 20 000 tonnes/year
- Lead–zinc slag: 100 000 tonnes/year
- Tin slag: 60 000 tonnes/year.

In addition smaller quantities of other slags are available from the refining of less common metals. The production of all non-ferrous slags is very localized but, in the vicinity of the works, they are an obvious supply of what might be expected by analogy with blast-furnace slag to be good roadmaking materials.

Uses in road construction

Little is known of the potential uses of these slags in road construction and the amounts produced in this country are too small to justify much research on their properties. They are produced in very much larger quantities in the USA and have been used in road construction to a limited degree (OECD 1977).

The main problem with such materials is that they may contain components that are unstable in the presence of water and/or that water leaching through them may contain harmful pollutants. An example of this occurred when slag from precious metal refining of silver and platinum was used as a base material for footpaths and car parks in the vicinity of the works. The silver and platinum slags were quite different in their properties and the silver slag gave no trouble. The platinum slag, on the other hand, caused failures which were attributed to the presence of sodium sulphide. This oxidizes to sodium sulphate and the oxidation is accompanied by considerable expansion. Soaking tests made on samples of the slag before it was used did not reveal any problems and tests for sulphate content had indicated very low values.

Imported slags

Most waste materials and industrial by-products are of low value and are used only in the areas where they occur. It is therefore extremely unlikely that any would be imported into this country from the continent. The only exception to this is the import of slags into south-east England which occurs for three reasons:

- the shortage of good-quality aggregates in that area
- slags, in general, are high-quality aggregates which can command a price high enough to make transporting them an economic proposition
- compared with the UK, slag production in neighbouring countries is much higher.

Slags produced on the continent do not differ substantially from those produced by similar processes in this country.

11. Slate waste

Occurrence

North Wales has always been by far the largest producing area of slate although it is also quarried in mid-Wales, Cornwall, Devon and Cumbria. Slate quarrying reached its peak in the 19th century when the development of the canal and then rail networks meant that slates could be transported cheaply from the slate-producing areas to all parts of the country for use as a roofing material. From the turn of the century slate quarrying fell into rapid decline as tiles once again became the cheapest method of roofing. During recent years there has been an increase in production but this is still at a very low level compared with Victorian times.

Only the rock suitable for splitting is acceptable because the main end use is for roofing. This factor, together with the losses incurred during cutting and splitting of large blocks of good slate, leads to a high proportion of waste at all stages; overall the proportion of waste-to-slate averages about 20:2. A comprehensive account of the production and uses of slate waste has been published by Watson (1980).

In 1991 the total annual production of slate waste was estimated to be 5–7 million tonnes (Whitbread *et al.* 1991). This was being added to the estimated stockpile of about 440 million tonnes, which puts it second only to colliery spoil in terms of the amount of waste available. However, unlike colliery spoil the stockpiles are highly concentrated in small, remote areas of the country (Table 33).

Table 33. *Location of slate waste stockpiles* (*Whitbread* et al. *1991*)

Location	Stockpile (million tonnes)
North Wales	350
Elsewhere in Wales	15
Lake District	15–20
Cornwall/Devon	5–10
Scotland	50
Total	435–445

Table 34. Composition of slate waste (Crockett 1975)

Component	Composition (%)
Sericite	38–40
Quartz	31–45
Chlorite	6–18
Haematite	3–6
Rutile	1–2.5

Composition

The nature of the slate waste varies according to its origin. Mill waste, mainly the ends of slate blocks and the chippings from the dressing of the slate, consists mainly of slate itself, but rocks such as cherts, which are sometimes interbedded with the slate, and igneous rocks may also be found in it. Slate, which forms the major portion of the waste, is a fine-grained aggregate of chlorite, sericite, quartz, haematite and rutile (see Table 34).

A typical chemical analysis of slate waste is given in Table 35. The waste is chemically inert and is most unlikely to cause any chemical problems when used in road construction. A summary of the physical properties of Welsh slate waste is given in Table 36. This shows that slate waste, provided that it can be crushed to satisfy the relevant grading requirements, will satisfy most other requirements with regard to factors such as plasticity, particle strength and durability.

Uses in road construction

Slate waste is, in effect, a crushed rock and is therefore potentially suitable for all applications where crushed rock is specified. Evidence for this is provided by the fact that it is used as a Type 1 granular sub-base material in those areas of North Wales where it occurs (Mears 1975) (see Fig. 43). It is thus apparent that it could also be used for bulk fill, as a selected granular fill material and as a granular capping material. With further processing and screening it would have wider use as an aggregate provided

Table 35. Typical chemical analysis of slate waste (Crockett 1975)

Component	Composition (%)
SiO_2	45–65
Al_2O_3	11–25
Fe_2O_3	0.5–5.7
K_2O	1–6
Na_2O	1–4
MgO	2–7
TiO_2	1–2

Table 36. Physical properties of slate waste aggregates (Goulden 1992)

Property	Source					
	Penrhyn	Ffestiniog	Llechwedd	Croes-y-Ddu-Afon	Aberllefni	Burlington
Water Absorption (%)	0.2	0.3	0.3	0.3	0.2	0.3
Flakiness Index (mean)	93	100	100	100	93	98
Elongation Index (mean)	23	29	34	34	23	27
ACV[a] (kN)	25	29	26	30	24	23
TFV (dry) (kN)	160	130	140	120	170	160
TFV (soaked) (kN)	110	90	80	70	110	100
AIV[b]	27	29	29	33	28	28
Relative densities						
Oven-dry	2.80	2.76	2.77	2.75	2.80	2.80
Sat. Surface Dry	2.82	2.78	2.78	2.77	2.82	2.81
Apparent	2.84	2.79	2.80	2.79	2.84	2.83
$MgSO_4$ soundness	99	98	98	98	99	98
Plasticity Index	0	0	0	0	0	0
Slake Durability Index (%)	96	94	95	94	96	96
Sulphate content (g.SO_3/litre)	0.01	0.01	0.01	0.01	0.01	0.01

[a] Aggregate crushing value
[b] Aggregate inpact value

Fig. 43. Processing of slate waste to produce granular sub-base

the usage did not have restrictions on the flakiness which is characteristic of the material.

Within its production area slate waste is used to manufacture a wide range of aggregates for use in road building and ancillary works and the product dominates the local market for road aggregates. However, the annual amounts used average out at only 250 000 tonnes, which is less than 5% of the current production of waste. Slate waste as a sub-base material has been shipped by sea to south-east England on a trial basis but, although the material proved to be satisfactory, the freight costs were marginally too high for it to be competitive.

12. Spent oil shale

Occurrence

Spent oil shale is the waste material from the now defunct oil extraction industry which was concentrated in the West Lothian area of central Scotland. At its peak in 1913 the industry employed 13 000 men and produced 3.3 million tonnes per year of waste (Burns 1978). Production steadily fell and it became increasingly difficult for the industry to compete with imported oil from the Middle East.

Production finally halted in 1962 by which time a large stockpile of spent shale had accumulated. This was estimated (Department of the Environment 1991a) at 100 million tonnes occupying 395 hectares of land in a small area around Livingston and Bathgate, about 20 km to the west of Edinburgh (see Fig. 44).

Fig. 44 Spent oil shale deposits, Livingston, Scotland

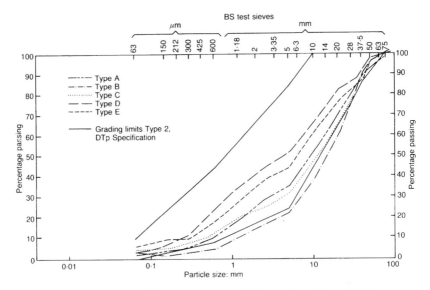

Fig. 45. Particle size distribution of spent oil shale samples (Burns 1978)

Composition

Following the extraction of the crude oil and naphtha from the oil shale, the spent shale, together with other materials considered unsuitable for processing, was deposited in heaps on land adjacent to the refineries and mines. Spontaneous combustion within the tips sometimes caused further changes to occur.

Although of different origin, spent oil shale is not unlike burnt colliery spoil in its chemical and physical properties. Like burnt colliery spoil it is pinkish in appearance and samples may be obtained with a particle size distribution corresponding to that required for granular sub-bases (Fig. 45). A typical analysis, compared with an analysis of burnt colliery

Table 37. Comparison of the chemical composition of burnt colliery spoil and spent oil shale

Component	Burnt colliery spoil[a] (%)	Spent oil shale[b] (%)
SiO_2	45–60	48.5
Al_2O_3	21–31	25.2
Fe_2O_3	4–13	12.1
CaO	0.5–6	5.3
MgO	1–3	2.2
Na_2O	0.2–0.6	NR
K_2O	2–3.5	NR
SO_3	0.1–5	3.2
Loss on ignition	2–6	3.0

[a] Range of values (Sherwood 1987)
[b] Typical analysis (Burns 1978)

Table 38. Results of sulphate and loss on ignition determinations on selected spent oil shales (Burns 1978)

Sample No.	Total sulphate (% as SO_3)	Loss on ignition (%)
1	2.20	1.87
2	2.80	1.25
3	0.70	1.35
4	2.40	2.26
5	2.80	1.07

spoil, is given in Table 37. The results of sulphate and loss on ignition determinations of a range of spent oil shales are given in Table 38.

The range of sulphate contents found in spent oil shale is much the same as the range found in burnt colliery spoil (Tables 37 and 38). For this reason the chemical problems associated with its use are also very similar and the earlier discussion on this issue of colliery spoil applies equally to spent oil shale. Spontaneous combustion is clearly not a problem because the shale was heated to extract the oil. Neither is the presence of sulphides because these will have been driven off or converted to sulphates during the extraction process.

Uses of spent oil shale in road construction

Due to its close similarity to burnt colliery spoil, spent oil shale may be used for all purposes where burnt colliery spoil is permitted to be used. It has been widely used as a bulk fill material in Central Scotland with very good results (Fraser and Lake 1967) and this is the main outlet for its use in road construction. Figure 45 shows that its particle size distribution may meet the grading requirements for granular sub-base materials and the Specification for Highway Works permits its use. However, as in the case of burnt colliery spoil, the imposition of particle strength and soundness requirements makes it unlikely that it would conform to all the requirements of the Specification.

13. Conclusions

Suitability of alternative materials

The review of the use of wastes and by-products in road construction in Part 2 has shown that all the mineral wastes and by-products currently available in the UK have some potential uses and that specifications do not unreasonably restrict their use. A survey by Mallett *et al.* (1997) of the use of alternative materials in roadmaking showed that all the materials considered in Part 2 had been used albeit on a very limited scale in some instances. The results of their survey are summarized in Table 39.

Table 39. Results of a survey of the use of alternative materials in UK road schemes (Mallett et al. 1997)

Material	No. of schemes identified	Road layer[a] in which the material was frequently used
Reclaimed bituminous material	146	Srf, SB, CA, EA
Shallow cold-mix in-situ	101	Srf, SB
Deep cold-mix in-situ	93	Srf, SB
Crushed concrete	55	SB, CA
Pulverized fuel ash	50	EA
Steel slag	40	Srf
Blast-furnace slag	36	Srf, RB, SB, CA
Colliery spoil	23	EA, CA
Crushed demolition material	11	SB, CA
Shallow hot-mix in-situ	9	Srf
Hot-mix off site	7	Srf, RB
Slate	7	RB, SB
Phosphoric slag	7	RB, SB
Tyres	7	Filter drains
China clay sand	6	RB
Cold-mix off site	5	Srf, RB, SB
Crushed brick	4	SB, CA, EA
Rail ballast	2	SB, EA
Furnace bottom ash	1	CA, EA
Glass	1	Srf
Spent oil shale	1	EA

Key for road layer: Srf, surfacing; RB, road base; SB, sub-base; CA, capping; EA, earthworks

Table 40. Summary of potential uses of alternative materials in road construction

Material	Bulk fill	Unbound capping layer	Unbound sub-base	Cement-bound material	Concrete aggregate or additive	Bitumen-bound material	Surface dressing aggregate
Crushed concrete	High[a]	High	High	High	High	Some	None
Asphalt planings	High[a]	High	High	Low	None	High	None
Demolition waste	High	Some	Some	Low	Low	Low	None
Blast-furnace slag	High[a]	High	High	High	High	High	High
Steel slag	Low	Low	Low	Low	Low	Some	High
Burnt colliery spoil	High	High	Some	High	Low	Low	None
Unburnt colliery spoil	High	Low	None	Some	None	None	None
Spent oil shale	High	High	Some	High	Low	Low	None
PFA	High	Low	Low	High	High	None[b]	None
FBA	High	Some	Some	High	Some	Low	None
China clay sand	High	High	Some	High	High	Some	Low
Slate waste	High	High	High	Some	Some	Low[b]	None
MSWI	High	Some	Some	None	None	None	None

[a] Suitable but inappropriate (wasteful) use
[b] PFA and slate dust can be used as a filler

There are few surprises from the results of the survey of Mallet *et al.* and the results largely confirm the conclusions of an OECD (1997) review. This identified reclaimed bituminous pavements, crushed concrete, blast-furnace slag, steel slag and PFA as being what the OECD review termed 'winners'. However, it is important to remember that the relatively lowly status of china clay, slate and spent oil shale is a reflection on the fact that they are not widely distributed and this does not reflect their potential. This is much better indicated by Table 40 which

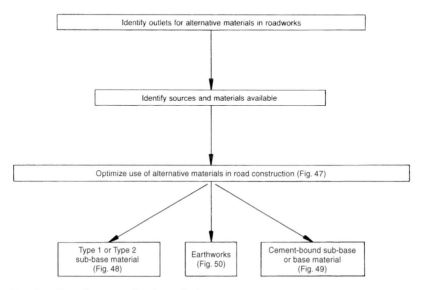

Fig. 46. Flow diagram: deciding whether to use an alternative material (based on BS 6543: 1985)

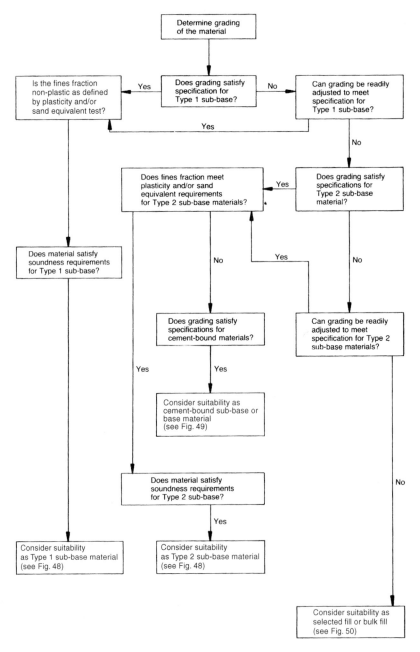

Fig. 47. Optimizing the use of alternative materials in road construction (BS 6543: 1985)

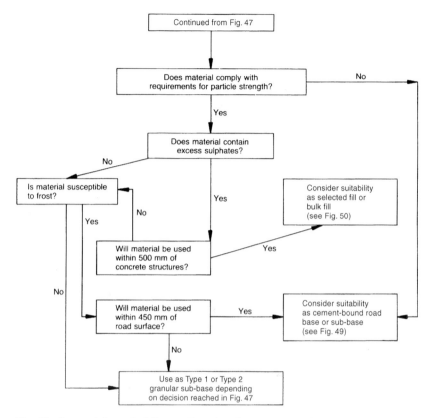

Fig. 48. Determining suitability as Type 1 and Type 2 granular sub-base (BS 6543: 1985)

summarizes the conclusions that can be drawn with regard to the materials considered.

Materials selection

The environmental aspects of deciding whether or not to use an alternative to naturally-occurring materials are considered in Part 3. However, assuming that an alternative material is readily available at economic cost, a decision still has to be made on whether or not it is suitable. The information given in Parts 1 and 2 should enable a decision to be made and this information is summarized in the flow diagrams given in Figs 46 to 50.

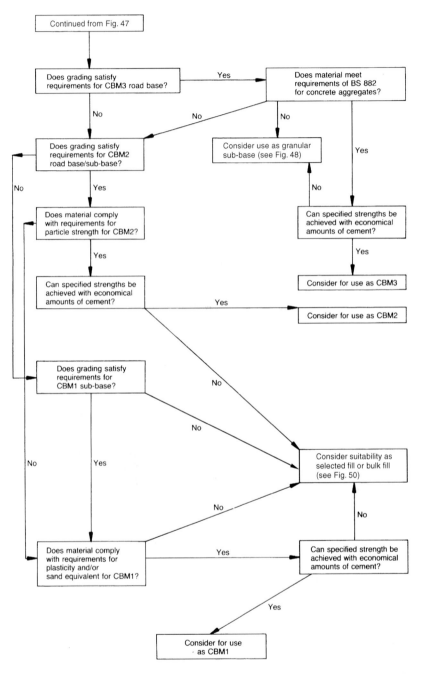

Fig. 49. Determining suitability as cement-bound road base or sub-base material

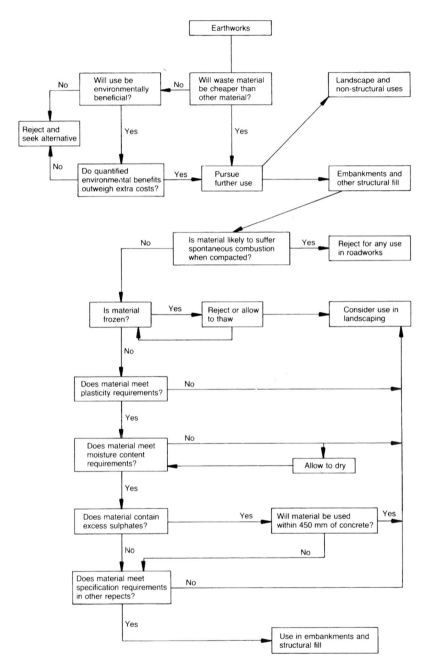

Fig. 50. Determining the suitability for use in earthwork construction (BS 6543: 1985)

PART 3
ENVIRONMENTAL AND
ECONOMIC
CONSIDERATIONS

Before any new road is built a cost-benefit analysis will have been made to ensure that the anticipated benefits to be gained exceed the economic costs and the environmental disturbance resulting from its construction. There is much scope for argument on this subject as can be seen from the prolonged and heated discussions that take place at public inquiries on the routing of new roads. No satisfactory method of assessment has yet been developed that will satisfy all parties on whether or not a road should be built. However, assuming that a decision has been made to build a road, all would agree that it is axiomatic that the materials used for its construction should be obtained at the lowest cost (commensurate with satisfactory performance) and with the minimum of environmental disturbance.

The economic factors relating to the materials used in construction mainly derive from the costs of extracting the material, processing and hauling it to the site, all of which are closely related to energy consumption. The environmental disturbance is made up of factors such as disturbance to the landscape leading to possible dereliction, disturbance caused by transporting material from source to site, and the depletion of natural resources. The three factors of resource depletion, environmental degradation and energy consumption are not completely independent of each other and steps taken to reduce the effect of one may be accompanied by an increase in another, but clearly any project should be planned in such a way that all effects are minimized.

The use of alternative materials, such as those considered in this book, in place of naturally-occurring materials, can help to achieve this aim. Care needs to be taken as the use of an unsuitable material for a particular set of circumstances can lead to a costly failure and create a climate of opinion that is hostile to its use. The primary requirement in selecting a waste material or by-product must, therefore, be that its use in preference to other materials will not shorten the life of the road structure. However, over-specification is wasteful of resources and a balance has to be struck between these two conflicting aims.

Parts 1 and 2 of this book dealt with the materials that are available and the uses that can be made of them in road construction. However, if a material is to be used the economic and environmental factors arising from its use also need to be considered. Part 3 is therefore concerned with a discussion of the factors that should be taken into account in deciding whether or not to use alternative materials, such as waste materials or by-products, in preference to naturally-occurring materials. It also describes the legislation relating to the use of waste materials and by-products in the UK.

14. Environmental effects of aggregate and waste production

Surface mineral workings are ubiquitous and, as can be seen from Table 41, occur throughout the country. The advantages of this are that a source of primary aggregate is never very far away and the haulage distances are therefore greatly reduced. The disadvantages are that large numbers of people are adversely affected and the environmental impacts of aggregates extraction are a source of increasing concern in many parts of the country. These impacts have been itemized (CPRE 1993) as loss of mature countryside, visual intrusion, heavy lorry traffic on unsuitable roads, noise dust and blasting vibration. The extraction of aggregates also represents the loss of two finite resources, the aggregates themselves and unspoilt countryside from which they are extracted. The harmful effects of aggregate construction can be considerably ameliorated by the attachment of restoration conditions to the planning consents so that it can be difficult to discern that any disturbance has occurred (compare Figs 51 and 52). However, the Department of the Environment (1991) admits that 'landscape conditions attached to surface mineral workings have often been ignored ... and the conditions are neither always implemented nor enforced'.

In the 20-year period to 1990 the total annual production within the UK of aggregates (sand, gravel and crushed rock) increased from 200 million tonnes to nearly 300 million tonnes. It was expected to continue to increase at a steady rate and the Department of the Environment (1994b) published forecasts that at a conservative estimate it would reach about 400 million tonnes/year by 2011. In fact the 300 million tonne figure for 1989 proved to be the peak year of production, by 1999 it had fallen to just over 200 million tonnes and was fairly stable at this level (Fig. 1). Even so the figures are still very large and in volumetric terms the amount of aggregates excavated has been likened (Adams 1991) to the equivalent of digging a hole 3.3 times the area of Berkshire $(1.25 \times 10^9 \, m^2)$ to a depth of 6 m in the period 1990–2010.

Road building plays a significant role in the demand for aggregates as it accounts for about one-third of the total production (Fig. 2). In 1989, 96 million tonnes were used and, even if the amounts used by local authorities in local road construction were excluded, it was estimated that the road building plans of the Department of Transport in the 1990s would use 510 million tonnes. This led the Royal Commission on Environmental Pollution (1994) to comment that 'We are concerned that extensive

Table 41. Location of surface mineral workings in the UK (Source: Quarry Products Association 2001)

Region	No. of Quarries	Annual production (million tonnes)
North	117	12
Yorks./Humberside	107	18
East Midlands	165	38
East Anglia	105	7
South East	285	32
South West	200	27
West Midlands	96	15
North West	66	15
Wales	109	23
Scotland	300	33

damage to the environment would be caused through extraction of the aggregates to carry out the present road building programme. We do not consider that the implied rate of consumption can be regarded as sustainable'.

The increasing concern about the harmful environmental effects of the proposed road construction programme which was reflected in the comments of the Royal Commission led the Government to review its policy and in 1998 it came up with what it claimed was a new, environmentally and economically sustainable roads policy. The main thrust of the policy was to:

Fig. 51. Gravel workings, Harlington, Middlesex (1990)

Fig. 52. Restored gravel workings, Harlington, Middlesex (1991)

- end the discredited 'predict and provide' approach which encouraged traffic growth by expanding road space to meet demand without sufficient regard to wider considerations
- move away from an exclusive reliance on road building and place much greater emphasis on finding other, more sustainable ways of tackling traffic problems
- move trunk road planning into the regional planning system so that strategic management and improvement is carried out in the context of the overall transport and land use strategy for each region
- put in place a better, more quality-sensitive way of evaluating road building proposals, against the key criteria of accessibility, safety, economy, environment and integration
- make things better for road users by giving road maintenance and traffic management a higher priority. The Highways Agency will be refocussed away from road building and towards strategic management of the network, making better use of what already exists.

However, even if no new construction took place there would still be a considerable demand for aggregates simply to maintain the existing road network.

Raw materials for civil engineering construction have to be obtained by opencast methods which scar the landscape (Fig. 51) while the mineral wastes for which there is little demand are usually stockpiled in spoil tips (Figs 9 and 13 give examples). In some cases the waste materials can be tipped into the holes produced by mineral excavations; this method is widely used for the disposal of domestic refuse. When this can be done it offers an attractive solution but, unfortunately, it is not

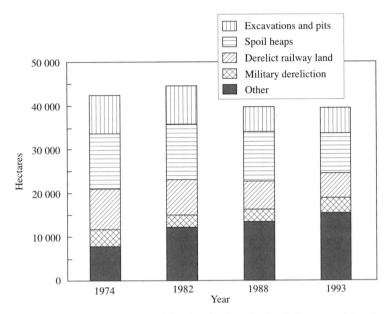

Fig. 53 Changes in the amount of derelict land in England, by type of dereliction (Source: The Environment Agency 2000)

always possible to do so. Apart from the fact that tipping into holes can cause pollution of underground water supplies, most waste materials are not produced sufficiently close to holes in the ground for such disposal to be an economic proposition. Consequently, the UK has large areas of derelict land made up of pits and waste heaps (Fig. 53). The latest estimate of the total amount of derelict land for which figures are available is more than 40 000 hectares, i.e. roughly the area of the Isle of Wight.

The amount of derelict land is, in fact, greater than is indicated by Fig. 53 because it only includes land which is defined as being 'so damaged by industrial or other development that it is incapable of beneficial use without treatment'. It does not include active workings, land to which restoration conditions have already been attached, or land which, although originally derelict, is no longer unsightly.

These exclusions go a long way to explaining the discrepancies between the official figures for the extent of the problem and the generally held view that there is much more derelict land about than the official figures suggest. From a layman's viewpoint the general definition suggested by the Civic Trust is probably better: 'Virtually any land which is ugly or unattractive in appearance: spoil heaps, scrap or rubbish dumps, excavations, dilapidated buildings, subsided or any land which is neglected'. To a casual observer the problem is enlarged by the fact that once a certain proportion of land in a given area becomes derelict the land in between, although not derelict by any of the definitions given above, appears to be so – a series of heaps or holes in an area blights the interstices as well.

Parts 1 and 2 of this book have shown that the alternative materials considered here can, in many instances, be used as substitutes to the naturally-occurring materials used in civil engineering. These parts were concerned with technical considerations but it is clear that their use can have considerable environmental benefits. These benefits are threefold as they can lead to the conservation of natural resources, the disposal of the waste materials, which are often the cause of unsightliness and dereliction, and the clearance of valuable land for other uses.

This fact has been recognized by the Government, which in 1992 stated and has since reiterated with increasing emphasis that: 'To reduce the environmental effects of quarrying new materials the Government is keen to encourage the greatest possible use of waste and recycled materials in accordance with the principles of sustainable development' (DOE/DOT 1992), sustainable development being defined (Department of the Environment 1994a) as meeting the needs of the present generation without compromising the ability of future generations to meet their own needs. The implications of sustainable development for minerals planning are that avoidable and irretrievable losses of natural resources should be limited. This means making the best and most efficient use of all available resources. It is thus clear that alternative materials, many of which occur in large quantities, can play an important role in achieving this aim.

Unfortunately the environmental benefits may be offset by increased economic costs and, even if an alternative material is technically suitable, there may be other considerations which mean that it is unlikely to be used. The problem was concisely summarized by the Advisory Committee on Aggregates (1976) in its reference to china clay sand:

'China clay sands need only the same basic grading and washing processes that are applied to natural aggregates. Much is readily available in heaps that are largely free of contamination by overburden and other deleterious matter. It is there virtually for the taking. To use it would conserve resources in other parts of the country, and permit land which might be used for aggregates construction to be used for other purposes. The continued existence of spoil heaps imposes costs on producers and the community alike – sterilizing further mineral development, destroying agricultural land and causing visual and other disbenefits. However, the material is located far from main centres of demand and the cost of transport by rail or sea has not proved viable. Thus we have the apparently nonsensical position that holes are being dug in one part of the country and heaps of sand being raised in another. The problem can only be solved on a national basis and the gap can only be closed by a national subsidy justified for environmental reasons. It is a sad commentary on our value judgement that the problem clearly has a lower priority than Concorde'.

(Note: the Concorde project involved the expenditure of £2000 million (sic), at 1976 prices; only 15 aircraft ever flew commercially and even these had to be given away because no airline would buy them.)

The dilemma remains; how these opposing factors can be balanced so that an optimum decision can be reached forms the subject of the remainder of Part 3 of this book.

15. Decision making – traditional or alternative materials?

The advantages to be gained by using alternative materials in place of naturally-occurring materials are finely balanced and it is rarely self-evident as to which is the optimum solution. No generally accepted method of making the decision has evolved but this chapter discusses the factors that have to be taken into account; it is based on a methodology originally put forward by Sherwood and Roe (1974) and further developed by the OECD (1977).

Advantages of using alternative materials
Conservation of natural aggregates
The use of waste materials for the construction of those layers where it is possible to use them can have significant environmental benefits and conserve the better quality materials for applications where they are really needed. Conservation of high-quality materials in this way will give greater benefit through long-term future use than is foregone by not using them now.

Benefits of removal of waste tips
The use of a tip as a source of material for road construction is an attractive proposition, but it is seldom feasible to remove the whole of the tip. The limited removal of material may hinder rather than help eventual restoration; if only partial removal of a deposit is planned then the stability and landscaping of the remainder will require consideration to avoid aggravating dereliction.

It is important to bear in mind that finding a beneficial use for the material in a tip is not the only satisfactory solution as it may also be re-contoured and vegetation established on its slopes, as in Fig. 54, so that it is no longer an eyesore and blends with the local landscape. This 'cosmetic' treatment has been widely used for reclaiming tips and it offers the best solution for dealing with the majority of tips. Even if all the bulk fill in new roads were composed entirely of wastes, the amounts used would be insignificant compared with the stockpile of wastes that is potentially available.

The costs of rehabilitating a tip by landscaping and planting can be fairly accurately assessed, so it is possible to quantify the benefit to be gained by using wastes in road construction by taking it to be equivalent to the cost of cosmetic treatment of the tip by landscaping and planting.

Fig. 54. Example of a reclaimed colliery spoil tip

Avoidance of borrow pits

As discussed in Parts 1 and 2, aggregates for the construction of the upper layers of a road pavement are almost invariably obtained from commercial quarries and gravel pits. However, for the lower layers, particularly for bulk fill, borrow pits are often used as the source of construction material (Fig. 55).

A borrow pit can be sited close to the road project so that haulage distances and the use by construction traffic of public roads are minimized. (It also allows the contractor to use unlicensed vehicles if haulage routes can be arrayed to avoid public roads. This point is often quoted in favour of borrow pits but is largely illusory. If the contractor does pay the tax this sum will be added to the contract so that, although the Government gets the benefit of the tax, it pays an equivalent extra sum for the road to be built. The sums involved are, in any case, small.)

The advantages to the contractor of using material from a borrow pit have tended to obscure the very real disbenefit that may result from the opening of a pit. Proponents of the use of wastes emphasize the dereliction that a borrow pit can cause, while at the other extreme claims are made that strict enforcement of planning laws would make it impossible for any permanent harmful effects to occur. In fact, although both extremes can occur, the situation lies somewhere in between. Whether or not a borrow pit will permanently disfigure the landscape depends partly on planning regulations and partly on topography. The restoration of borrow pits in undulating countryside can be carried out comparatively easily by landscaping so

Fig. 55. Borrow pit opened to supply fill material for M4/M25 interchange (1984)

that although the topography is affected, the change is not apparent to any-body unacquainted with the area before the operations began. This form of restoration can be carried out quickly and leaves no permanent scars.

In flat country, however, particularly where the water table is high, landscaping may not be a practicable proposition. If the pit is to be rein-stated, inert filling materials, which may not be readily available, have to be imported to the site (if they were readily available they could well have been used instead of the borrow material). If the pit is left as a lagoon the land is permanently lost and, although it may be possible to use the lagoon for water sports, the demand for such uses is limited.

An examination of the after-use of the borrow pits opened up for the eastern sections of the M3 and M4 motorways (Sherwood and Roe 1974) showed that, whereas all the pits opened in undulating country had been effectively restored, many of the pits in the flat areas were lying derelict 15 years after the borrow pits had been opened, a situation that still applies after nearly 40 years (Fig. 56).

Disadvantages of alternative materials

Increased haulage costs

The use of wastes rather than borrow material usually involves additional haulage costs because a borrow pit will, almost invariably, be closer to the road site than a waste tip (if it is not, the advantage of using waste is so overwhelming as to be self-evident). In some cases it will be necessary to strengthen the roads used for haulage, and the costs of doing so should be added to the extra costs of haulage.

Fig. 56. Site of a borrow pit opened for the M4 motorway in 1962 and still derelict in 2001

Disturbance caused by haulage

The disturbance caused by haulage is an environmental disbenefit as the haulage lorries use public highways causing congestion, noise, dust and deposition of material on the road. As this congestion is not easily quantified it tends to be ignored, but it is nonetheless real to people who live near the haulage route. The removal of material from a spoil heap can easily mean the exchange of a permanent, but quiescent, disbenefit, to which local inhabitants have to some extent become accustomed, for a widespread mobile nuisance.

No generally agreed procedure has been developed for quantifying traffic disturbance but, in making a decision on whether wastes should be used, assessments can be made of the length of the haulage route, the volume of traffic generated and the number of people likely to be affected by traffic.

Greater variability of waste materials

All of the waste materials considered in this book have at least some potential for use in road construction as substitutes for naturally-occurring materials, particularly for bulk fill. Most differ in their physical properties and chemical properties to the materials normally used but this does not necessarily place them at a disadvantage.

Particular attention has to be paid to their chemical properties which may be highly variable. Some, notably china clay sand and slate waste, are completely inert chemically and will present fewer problems than are likely to occur with natural materials. Others, such as blast-furnace slag and PFA, although different in chemical composition from the naturally-occurring materials used in road construction, are unlikely to prove to be any more difficult to use. They can present problems that are unique to the particular material but this does not impose any serious limitations on their use.

However, in the case of some waste materials unusual problems arising from their chemical composition do need to be considered. This does not mean that all such materials should be regarded with suspicion but it does mean that tests that would not normally be considered necessary have to be carried out to ensure that they are suitable for use. Concerns about the potential environmental effects of alternative materials are a barrier to their wider use. There are few well-established procedures and most cases are decided on a site-specific basis by the regulatory authorities. This leads to uncertainty, delay and increased costs, with the result that contractors are reluctant to propose the use of these materials.

In considering the use of wastes, care should therefore be taken not to regard them as being equivalent in all respects to the natural materials they are expected to replace; this is rarely the case. Aggregates produced by commercial suppliers are processed to meet the requirements of a particular specification. The user can reject the material if it fails to meet the specification requirements and can require the supplier to rectify any problem that occurs. Although waste materials can be processed in the same way as natural aggregates this greatly increases the costs and cannot be justified unless large quantities are being regularly taken from the stockpile. This means that they require more testing on site than do commercially produced aggregates.

Possibility of long-term pollution problems
Naturally-occurring materials used in road construction are most unlikely to give rise to any problems of pollution arising from the leaching of soluble materials into watercourses. Most are chemically inert and contain insignificant amounts of soluble material that could be removed by leaching. The very fact that they occur naturally gives confidence in their use because any soluble materials that they might contain would have given rise to problems in the areas in which they occur. If such problems had arisen, this fact would have been realized so that their use could be avoided.

This is also true of china clay wastes and slate wastes that are naturally-occurring quarried materials. It is not true, however, of most of the other materials considered in this book. Many of these contain soluble materials that could be a potential problem, and this fact is considered in a subsequent chapter.

Conclusions
The preceding sections showed that the use of alternative materials involves a benefit (A) usually as a result of removing a waste tip, but also involves a disbenefit made up of transport costs (C), the disturbance caused by haulage traffic (D), the extra costs (E) arising from the greater

variability of waste materials and a factor (F) arising from the greater potential for water pollution if waste materials are used. Therefore, as it is a fundamental concept that the money spent on controlling a disbenefit should not exceed the cost that the disbenefit is causing, the relation between A, C, D, E and F can be written in the form

$$A \geq C + D + E + F$$

However, if waste material is not used the material has to be obtained from a borrow pit which involves a disbenefit (B). Therefore the above equation has to be modified to

$$A + B \geq C + D + E + F$$

This relation is deceptively simple and if all the factors could be measured in the same units there would be no problem in deciding when to use wastes and when to use borrow pits. As the above discussion has shown, only A, C and, possibly, E of the factors to consider can be measured in monetary terms; even if an assessment is made of B and D the results are not in the same units. How, for example, can the disbenefits caused by borrow pits, which may be permanent, be compared with the temporary disturbance caused by traffic, particularly as the people suffering from the effects of one may not be the same group of people affected by the other.

The answer is that somebody has to make a subjective judgement, but in doing so all the factors have to be given due consideration. As the decision is subjective it will not generally prove possible to convince everybody that the correct decision has been reached. For this reason it is necessary to consider, in the absence of a generally accepted method of comparing all the factors involved, what further steps might be taken to improve the present position. Ways in which this might be done are therefore considered in the next chapter.

16. Encouraging the use of alternative materials

There are many environmental benefits to be gained by using alternative materials as substitutes for the materials that are traditionally used in road construction. However, if alternatives are to be so much as considered as potential substitutes, positive steps need to be taken to encourage their use. Possible methods by which this can be done are considered in this chapter.

Amendments to existing standards and specifications

The situation with regard to standards for alternative materials was considered in Part 1, which discussed the European Standards that are being prepared to cover the use of alternative materials. Good progress is being made with the preparation of these standards and, in the long-term, the relative absence of standards that give the requirements for alternative materials will be rectified.

A survey of the requirements and specifications for improved use of unbound granular materials (European Commission 2001) concluded:

- that the optimization of the use of marginal and alternative materials in pavement layers would require a coupled structural layer design and material selection
- there was a need to change design thinking from 'Where can I find an aggregate to meet my need?' to 'How can I beneficially use this available aggregate to build successfully?'
- trying to make alternative materials 'fit' into conventional aggregate roles will probably be either costly or unsuccessful
- specifications must be used which facilitate the use of alternative materials
- the desire to use scarce resources most efficiently and the consequent desire to use alternative materials is likely to lead to the increased use of performance-related and end-product specifications.

The European Commission report contains draft specification clauses along the lines suggested above but much more work remains to be done on this subject. In the meantime, the Specification for Highway Works gives valuable guidance on the use of many of the alternatives considered here. For example, it mentions by name many of the materials as being suitable for particular applications. This is true of blast-furnace slag and crushed concrete which are named as being suitable for nearly all the uses covered by this book. Where they are considered suitable, other materials such as colliery spoil and PFA are also mentioned by name but some materials, notably china clay sand and slate waste, are not mentioned at all. Absence of a specific mention does not, however, preclude use if a material fulfilled all the requirements for a particular use and it could be argued that it fell into a category that was permitted. For example, slate waste is a type of crushed rock, china clay sand differs little from quarried sands, and spent oil shale is very similar to burnt colliery spoil.

It can therefore be concluded that the Specification takes a permissive attitude to the use of wastes and, generally speaking, it would be difficult to argue that it is unduly restrictive. There are certain anomalies in the Specification relating to restrictions on particular materials which seem difficult to justify; these are considered in the relevant sections of the report. However, these anomalies are relatively minor and even if rectified would do little to add to the amount of waste materials already being used in road construction.

This conclusion is borne out by two critical appraisals for the Department of the Environment of the specifications for road and building construction in use in this country (Collins *et al.* 1993, Collins and Sherwood 1995). These confirmed that the Specification allowed a wide range of materials to be used in road construction. They also found that although all highway authorities accept the Specification as a basis for their own specifications, many make significant departures from it. These departures were, on the whole, likely to lead to more stringent requirements being imposed rather than to any relaxations.

Assessment of experience in other countries
The OECD (1977) reviewed the use of waste materials and by-products in road construction in member countries – Belgium, Canada, Denmark, Finland, France, Germany, Italy, the Netherlands, Spain, Switzerland, the UK and the USA took part. This review showed that most of the larger countries produced considerable quantities of colliery spoil, PFA, ferrous slags and demolition wastes, and all the other materials discussed in this book were produced in at least one other country. Some wastes discussed in the OECD report did not occur in the UK, while in the case of china clay sand the UK and the USA were the only producers.

For each waste and by-product judged by the OECD committee to have potential for use in road construction, a detailed discussion of the physical and chemical properties, established uses and the problems that might occur was given in the report. The review showed that, at that time, the UK was well to the fore in developing uses for waste materials and by-products in road construction and, for the most part, had more to offer by way of experience in their use than it could learn from experience in other countries.

Few of the materials available in the UK were being used in other countries for purposes for which those materials were not already being used or considered for use in this country. The major exception was the use of blast-furnace slag, particularly in France where, relative to air-cooled slag, granulated slag was produced in much larger quantities and found many outlets in road construction (as discussed in Chapter 10).

A later OECD review (1997) drew much the same conclusions as to the relative merits of the available alternative materials and produced a series of six overall recommendations designed to better enable recycling efforts to keep pace with increasing global stores of by-products and decreasing space for landfills. These recommendations were:

- test materials before recycling
- ensure that recycled materials are used wisely
- promote the increased use of proven recycling solutions
- balance regulations and policies that foster recycling and discourage dumping
- balance engineering, environmental and economic factors
- increase research and knowledge transfer.

A literature review undertaken as part of a collaborative research project funded by the European Commission (ALT-MAT 1999) showed that there was a general political will to encourage the use of alternative materials in all countries. This was expressed in various ways; sometimes as direct legislation, sometimes as action plans and directives. Most countries had set targets for increasing the amount of recycling. Most employed a landfill tax, and in some a tax on natural aggregates had been introduced or was being considered.

Economic factors, such as transport costs and treatment costs, still limited the use of alternative materials, particularly in countries with large reserves of natural aggregates. In densely populated urban areas, however, the use of alternative materials was increasingly becoming economic.

Technical specifications for road construction in most countries used the same tests for natural and alternative materials. The alternative materials were assessed on the basis of the natural materials they most closely resembled. In some countries, a distinction was made between road by-products, which may be recycled for the same application with minimal testing, and non-road by-products, for which a comprehensive testing programme was required.

Guidance on the use of alternative materials

One factor that could inhibit the greater use of alternative materials in road construction is the absence of guidance on their use. The Specification for Highway Works, together with the accompanying Notes for Guidance and the Advice Notes which are issued from time to time, advise on the construction methods to be employed with all roadmaking materials and give some guidance on the use of certain wastes. There is also a British Standard BS 6543 (1985) on the use of wastes in civil engineering, but knowledge of this standard does not seem to be widespread (Collins et al. 1993). Apart from these publications, producers of the major wastes publish technical information which is freely available, and symposia and conferences are held on the use of wastes. Moreover since the early 1990s the Department of the Environment (now the DTLR) has sponsored a great deal of research on the use of alternative materials and the results are promulgated in various official publications. To publicize the use of alternative materials it set up an Aggregates Advisory Service (AAS) the aim of which was to assist the Government in achieving its objectives of reducing the construction industry's dependence on land-won primary aggregates and increasing the contribution from secondary and recycled materials. The AAS issued leaflets, available for downloading on the internet, on the properties and uses of all the major materials available. The AAS was discontinued in 1999 but its work has been taken over by the Waste Management Information Bureau (WMIB) which is based at AEA Technology Ltd at Harwell.

The amount of research sponsored by Government departments and others has, with some justification, come in for criticism (Mallett *et al.* 1997) on the grounds that there is a degree of repetition (it is particularly ironic that there should be a recycling of research on a subject connected with recycling!), omission of research in some important areas and inefficient dissemination of research results. However, it is clear that guidance on the use of particular wastes is not lacking for those who wish to use them. More important is the attitude of prospective users of wastes. If they are not interested or have no incentive to use wastes, no matter how much information is published it will remain unread.

Changing attitudes to the use of alternative materials

The survey by Collins *et al.* (1993) did not reveal any widespread resistance to the use of alternative materials; they were not perceived to be of a lower grade, or more troublesome to use. The main reason for not using them was that there was rarely any incentive to do so. The naturally-occurring materials used in road construction are relatively cheap and readily available (perhaps too much so) and their cost forms only a small proportion of the total cost of the road. There is thus a tendency to 'play-safe', even if such materials do cost more; criticism of needless expenditure is muted compared with the criticism that occurs if a costly failure should arise through the use of a material that subsequently proves to be unsuitable.

Any environmental benefits to be gained by using lower-quality, but acceptable, materials are ignored because there is no provision for taking them into account. For this reason those responsible for the choice of material are likely to be influenced by a relevant maxim from another field that 'nobody was ever sacked for choosing IBM'. The use of alternative materials will thus remain a laudable aim which will not be fulfilled unless some means can be found of swinging the balance in favour of their use. How this might be done is considered next.

Administrative measures to increase the use of alternatives

In the absence of any generally accepted method of putting a monetary value on the environmental benefits to be gained from the use of wastes, other methods have to be used to encourage their use. Policy methods that have been put forward are discussed in the remainder of this section.

Nomination of the use of a particular source of waste material
The use of alternative materials as secondary aggregates rather than using naturally-occurring primary aggregate materials is such an attractive proposition that there has for long been much pressure on the successive Government departments and agencies responsible for road construction to utilize them in their road building programmes.

Although central Government now actively encourages the use of alternative materials in its road construction programmes, it is and has always been reluctant to insist that alternatives such as waste materials should be used, even when they are readily available. This stems from the fear that if it did so and the nominated source gave rise to unforeseen problems (either real or imaginary), it would be exposed to claims (possibly unreasonable) for compensation by the contractor. It has therefore preferred to specify a wide range of materials, which in many instances includes waste materials and by-products, for any given purpose. This then allows contractors complete freedom of choice of the material (provided it complies with the Specification for Highway Works) within this range. The former Department of Transport's past attitude on this subject had been widely criticized, not only by environmental pressure groups but also by official bodies such as the Advisory Committee on Aggregates (1976) and the Royal Commission on Environmental Pollution (1974 and 1994).

As can be seen from the frequent references to waste materials in the earlier editions of the Specification for Highway Works those responsible were not unfavourably disposed to the use of wastes. However, until recently there has been a reluctance to sponsor the use of wastes on the part of those responsible which can be summed up by a comment made by the Environment Committee of the House of Commons 'The majority of those involved with waste in this country appear to be guilty of thinking without imagination and planning without ambition, of finding problems instead of solutions and aiming at short-term goals without a vision of resource use and waste management which we should be striving for' (House of Commons 2001). Not being part of the solution they became part of the problem and, judging from the Environment Committee's comment, this is still true in the case of the disposal of household waste.

There was a time when this accusation could, with much justification, be levelled against the former Department of Transport and former Department of the Environment. These Departments have now been merged into the DTLR and its road building responsibilities hived off to the Highways Agency. The complete change in tone since the early 1990s means that it is no longer true and the Highways Agency now actively promotes the use of alternative materials in road construction. In this connection it has prepared the following procedure for evaluating new materials which is carried out in the following five stages.

- *Stage 1. Desk study.* To assess and evaluate existing information on the material.
- *Stage 2. Laboratory study.* Tests of the mechanical properties of materials to allow theoretical predictions to be made of their performance.
- *Stage 3. Pilot-scale trials.* Evaluation of construction and performance of materials in small-scale trials.
- *Stage 4. Full-scale trial.* Trial on a trunk road to establish whether the previous assessments obtained from Stages 2 and 3 are realized.
- *Stage 5. Highways Agency Specification Trials.* This stage is necessary to carry out further evaluation of the material and to test the specification under contract conditions.

Notes
Stages 1 to 4 are financed by the manufacturer of the material. For Stage 5 the additional cost, if any, of the material is borne by the manufacturer. Stages 1 to 4 can be carried out by the TRL or other independent organization. In the latter case, the reports are appraised by the TRL.
In all cases, the new materials are compared with conventional materials to obtain comparative performance.

Imposition of a tax on the use of naturally-occurring materials (aggregate levy)
A report commissioned by the Department of the Environment on the occurrence and utilization of waste materials (Whitbread *et al.* 1991) concluded that the most promising line of policy intervention would be to introduce a tax on the use of primary aggregates. Such a price rise for primary aggregates could have two significant impacts on the economic benefits of a given road proposal:

(*a*)   it could promote more careful scrutiny of the net economic benefits of the road proposal
(*b*)   it would encourage greater efficiency in use of primary aggregates and greater use of alternative materials.

The proposal was the subject of much debate and after prolonged discussions the Government decided to go ahead and introduce a tax on the quarrying of natural aggregates. In his budget statement of March 2000 the Chancellor of the Exchequer announced that a levy would be imposed on primary aggregates (both land-won and marine dredged) as from 1 April 2002. Secondary and recycled materials would be exempted in order to encourage their use. Part of the revenues would be recycled through a Sustainability Fund which would be used to deliver environmental benefits to local communities affected by quarrying.
This proposal followed research commissioned by the then DETR which, not surprisingly, showed that there are significant local environmental costs associated with the extraction and transport of aggregates, including noise, dust, vibration, loss of biodiversity and amenity and visual intrusion. It was not just local communities who suffered but

there was also evidence of wider public concern over the environmental impact of quarrying in protected areas such as national parks. The research found that the average environmental cost associated with the extraction and transport of aggregates was around £1.80 per tonne.

The Government declared that it was taking a cautious approach by introducing the levy at a lower rate than that justified by the research and by giving firms two years to plan for its introduction. The levy would apply at a rate of £1.60 per tonne to sand, gravel and crushed rock extracted in the UK or its territorial waters. To protect international competitiveness, the tax would be levied on imports but exports would be exempt as would recycled aggregates.

Predictably the quarrying industry expressed outrage at the proposals. It estimated that it would add £385 million to the costs of construction projects and claimed that it would bring no environmental benefits because it did not distinguish between responsible quarry operators and those that did nothing to improve their environmental performance (QPA 2001).

Imposition of a tax on the disposal of construction wastes to landfill (landfill tax)
This is a mirror image of the tax on the use of primary aggregates considered above. Until recently much material that has potential for use in construction work was simply dumped as landfill where, for example, it was used to reclaim land that has been used for aggregate excavations. Although the use of such material is not without benefit in land reclamation, it obviously makes more sense to use it rather than to create more holes in the ground to obtain primary aggregate.

The determination to reduce the amount of waste going to landfill and to encourage recovery, recycling and reduction led to the introduction, in October 1996, of a tax on the landfilling of waste. The tax was introduced to ensure that the price of landfill waste disposal reflected its environmental impact. Traditionally landfill has been a cheap disposal option and although costs have risen as more stringent requirements have been imposed it has remained the cheapest disposal option.

The tax was introduced at a rate of £7/tonne for active wastes and a lower rate of £2.00/tonne for inactive wastes and it now has an important role to play in the Government's overall strategy for promoting more sustainable waste management. In the 1999 Budget the tax was increased to £10/tonne rising by £1/tonne each year to £15/tonne by 2004. The Government hopes that this will send a strong price signal to waste managers, including those dealing with municipal waste, to reduce their dependence on landfilling. It builds on the existing role of the tax to encourage waste minimization, re-use, recycling/composting, and recovery of waste. The Government's review of the waste strategy, planned for 2004, will provide an opportunity to assess the effect of the tax increases in light of experience.

The landfill tax credit scheme allows up to 20% of the funds generated by the tax to be channelled into bodies with environmental objectives. The aims of this scheme are:

- to help promote and foster sustainable waste management practices which provide alternatives to landfill
- to help projects which benefit communities in the vicinity of landfill sites, thereby helping to compensate for the disamenity effects and environmental impacts of landfill.

Imposition of a requirement that all projects should incorporate a proportion of waste or recycled material

A less controversial approach than that of a tax on naturally-occurring materials and which does not have the disadvantage of the above nomination approach is to impose a general requirement that a given proportion of the materials used in a construction project should be either waste or recycled material. This is the approach that has been adopted in some countries, e.g. Germany and the Netherlands. It is backed up in Germany by a law which prohibits the disposal of mineral wastes which are capable of being recycled.

Such requirements have aroused interest in the use of waste and recycled materials and may well mean that more of them are used than the law demands. This also means that standards and specifications are produced for their use that mirror those already in existence for naturally-occurring materials. Sweere (1989) reported that in the Netherlands half of the market for granular sub-base materials is made up of recycled construction and demolition waste. The granular materials are produced in special plants in a process of crushing and cleaning. The cleaning involves the magnetic removal of iron and the removal of wood, plastics, etc., through a washing process. Sweere (1991) also made reference to Dutch standards that give the requirements for recycled materials.

Reservation of high-quality materials for high-grade use

This is a mirror image of the imposition of a requirement that all projects should incorporate alternative materials. It would reserve higher-grade materials for those purposes for which the scarce properties of the material were essential. Thus, valley gravels which could be washed and used as high-grade aggregates would be used for ready-mixed concrete where alternative materials are not readily available, but never in an unbound form where numerous alternatives exist.

In accordance with the principles of sustainable development it is clear that aggregates should be used for the highest grade applications for which they are suited. This would not only conserve high-grade aggregates for high-grade use but progressively encourage aggregate users to adjust their demands to make greater use of lower grade materials which would usually be cheaper. The CPRE (2000) suggests that conditions should be imposed requiring uses to be limited to those known to require material of the grade available, e.g. materials suitable for the surfacing of trunk roads. However, such a proposal would involve some form of end-use control over quarried materials and, apart from being unpopular with producers, this would be difficult to enforce.

17. Policy and controls on the supply and use of construction materials

In this chapter the main policy and control factors affecting both naturally-occurring and alternative materials are discussed. The aim is to give a background to official policy with regard to the controls that are imposed on the use of construction materials; it is not intended to be exhaustive as some aspects are complex and are further complicated by frequent additions and amendments.

The enormous amounts of materials (principally aggregates) and the effects that the supply and use of these may have on the environment inevitably mean that the Government has been forced to play an active role in ensuring that the naturally-occurring materials required by the construction industry are made available. At the same time the amounts of waste materials that are produced and the fact that some of these may be hazardous has also meant that Government action has had to be taken on the use and disposal of these products. Earlier parts of this book have shown that some of these waste materials can be used as alternatives to naturally-occurring materials. Policy and control on the supply and use of traditional aggregates and alternatives therefore become interlinked. Government policy on both of these issues was summarized in two papers (Department of the Environment 1994a and DETR 2000). These stated the following.

> With increasing demands for minerals the key issues for sustainability in this sector are:
>
> - to encourage prudent stewardship of mineral resources while monitoring necessary supplies
> - to reduce environmental impacts of minerals provision both during minerals extraction and when restoration has been achieved.
>
> The Government will therefore:
>
> - continue research on the availability of mineral resources and the environmental costs and benefits of using different sources
> - promote more efficient use of mineral resources in general, for example, encouraging recycling of materials and substitution of alternative materials, where appropriate.

Policy on the availability of construction materials
It is Government policy to ensure that there is an adequate supply of minerals to meet the needs of the construction industry. To achieve this

aim Regional Aggregates Working Parties (RAWPs) have been set up to prepare supply and demand forecasts for their particular region using data made available by the DOE of long-term forecasts for primary aggregates. The regional commentaries prepared by each RAWP form part of the Mineral Policy Guidance (MPG) Notes issued by the DTLR. These guidance notes set the parameters for minerals planning policy at national and regional level. The current edition of MPG 6 (Department of the Environment 1994b), *Guidelines for Aggregates Provision in England*, states that one of the aims of the Guidance Note is to 'provide guidance on how an adequate and steady supply of material to the construction industry, at a national, regional and local level, may be maintained at the best balance of social, environmental and economic cost, through full consideration of all resources and the principles of sustainable development'.

The 1994 edition of MPG 6 placed greater emphasis on sustainability than hitherto. It identified the potential conflict between increasing aggregate demand and the need to limit the impact of extraction industries on the natural environment. The Guide noted the importance of the use of secondary aggregates and recycled materials which it estimated accounted for 10% of the market in 1989 and set targets of 40 million tonnes per annum by 2001 and 55 million tonnes by 2006 (in 1989, 40 million tonnes would have represented 17% of the market).

Six months after the publication of MPG 6, the Royal Commission on Environmental Pollution published a report (1994), containing the recommendation that the proportion of recycled materials used in road construction should be doubled by 2005 and doubled again by 2015. The targets set by MPG 6 and the Royal Commission are not incompatible, but the realization of these targets would require planning policy changes to encourage alternative sources of construction materials to enter the market – how this might be done has already been considered in a previous chapter. The rate and use of recycled aggregates and other alternative materials is, in fact, already higher than was thought to be the case when MPG 6 was last reviewed in 1994 and the targets it set for 2001 have already been achieved.

It is anticipated that a new edition of MPG 6 will be published in 2002 and, as part of the consultation exercise, the Government declared that the new planning policy should properly reflect its principles for sustainable development. This means a significant shift away from the previous 'predict and provide' policy in which predictions were made about future demand and policies then put into place to meet this demand. Such a policy is incompatible with sustainable development, the forecasts became targets to be achieved and those on which the 1994 edition of MPG 6 were based were wildly inaccurate (in 1992 the lowest estimate for aggregate production in 2000 was 300 million tonnes whereas since 1996 it has been fairly stable at about 220 million tonnes/annum, see Fig. 1)

Key elements of the new strategy for the revised edition of MPG 6 are expected to be to:

- reduce the demand for aggregates through better design of construction, less waste of construction materials and development of alternative construction materials
- promote and maximize the use of recycled and reclaimed materials
- minimize the amounts of aggregates that need to be newly extracted.

Legislation concerning construction materials

Natural materials

The excavation of aggregates and fill materials is subject to planning controls exercised by the local planning authority (LPA). An LPA can refuse permission for excavation of aggregates but its ability to do so is limited and the applicant can always appeal to the DOE against the refusal. LPAs are required to prepare a local Minerals Plan which must incorporate the supply and demand figures prepared by the RAWPs. LPAs cannot challenge the national forecasts, only the amount of minerals they are expected to produce towards meeting total forecast demand.

In the case of borrow pits opened up to provide fill materials for road construction, the LPA has more scope because, generally speaking, the material from a borrow pit will not be classed as an aggregate. The arguments for and against the use of naturally-occurring materials from borrow pits rather than the use of waste materials was discussed earlier in Part 3. If planning consent is given by the LPA for the opening of a borrow pit it will invariably be subject to conditions requiring that the pit should be restored to some beneficial use.

As the opening of a borrow pit to obtain fill material is subject to its control, the LPA can to some extent (but not wholly, because there is a right of appeal to the DOE if a planning application is refused) play a part in deciding on the source and type of material that should be used for bulk fill. If suitable waste materials are available within a reasonable distance of the site the LPA would probably insist on their use and refuse permission for a borrow pit to be opened.

Waste materials

Controlled wastes.   The disposal of waste materials is covered by the Environmental Protection Act (DOE 1992). 'Controlled wastes', as defined by the Act, means household, industrial and commercial waste. Briefly, under the Act a material is defined as a waste if the answer to any of the following questions is 'yes'.

(*a*)   Is it what would ordinarily be described as a waste?
(*b*)   Is it a scrap material?
(*c*)   Is it an effluent or other unwanted surplus substance?
(*d*)   Does it require to be disposed of as broken, worn out, contaminated or otherwise spoiled?
(*e*)   Is it being discarded or dealt with as if it were a waste?

Waste from commerce or industry which falls into any of these categories is a 'controlled waste' and the producer of the waste has 'a duty of care' to ensure that its disposal does not cause any environmental problems. To ensure that it does not, if the waste leaves the premises on which it is produced the waste carrier must be registered with a waste regulation authority and transfer documentation will be required. Wastes to be treated as industrial wastes include:

(a)   waste arising from tunnelling or from any excavation
(b)   waste removed from land on which it has been previously deposited and any soil with which such wastes have been in contact.

At first sight it would seem, therefore, that all the materials discussed in this book fall within the controlled wastes covered by the Act. However, wastes from mining and quarrying operations are specifically excluded. The argument in favour of their exclusion is based on the fact that disposal in the mine or quarry falls within the terms of the original planning consent. This means that the Act does not apply to colliery spoil, china clay sand, slate waste and spent oil shale (see below). Nor does it apply to the current production of blast-furnace slag as it is produced as a manufactured product all of which is sold to the construction industry.

Hence only construction and demolition wastes, FBA and PFA fall into the category of controlled waste. But stockpiles of construction and demolition wastes are also exempted from the requirement to obtain a disposal licence 'provided that the deposit is made for the purposes of construction currently being undertaken on the land on which the waste is deposited (or will be used for future construction within three months)'. If any doubts exist about the classification of a particular material, the local Waste Management Authority should always be consulted.

Other wastes.   The fact that mining and quarry wastes are not classified as controlled wastes does not mean that there are no controls placed on them. Producers must apply for planning permission to the relevant minerals planning authority for all new spoil disposal schemes and, if approved, conditions are attached to the planning permission to safeguard the surrounding environment. However, some tipping still takes place on tips of long standing with no conditions attached, as permitted development under the General Development Order. The responsibility for enforcing the planning regulations rests with the LPA which, outside the metropolitan conurbations of England and Wales, is the appropriate County Council.

The safety of spoil tips is covered by legislation designed to control surface tipping rather than to eliminate it. Thus the Mines and Quarries (Tips) Act of 1969 was passed as a response to the Aberfan disaster of 1966 and was designed to ensure that spoil tips do not constitute a hazard to life and property. This Act has meant that the conical-shaped spoil tips of the past have largely disappeared to be replaced by tips with plateau-shaped surfaces (see Fig. 13) with slopes that are more stable and less likely to slip. Precautions are also taken to ensure that tips do not catch fire, with the result that the amount of burnt colliery spoil that is available is steadily declining.

Recycled materials.   Recycled materials can fall under the heading of controlled wastes but there is the added complication that recycling plants also need to have planning consent. A fixed-site recycling plant involving the import of materials and the export of an improved recycled product is an unwelcome neighbour. A planning consent may therefore be granted only after consideration of factors relating to both the site and the proposed operation. The final consent may be conditional and impose requirements to limit the impact of noise, visual intrusion and air pollution.

18. Health and safety considerations

Hazards to site personnel

None of the materials considered in this report is particularly harmful and, on the whole, their extraction and use requires no additional precautions over and above those normally required in civil engineering construction, which are covered by the Health and Safety at Work Act 1974 and earlier legislation. However, certain problems arise which would not normally be encountered with natural materials. The combustion of colliery spoil can give rise to high local levels of toxic gases such as carbon monoxide, sulphur dioxide and hydrogen sulphide which may be a hazard to workmen removing spoil where excavation exposes an area of burning material. The instability of seemingly hard surfaces of waste deposits may also be a problem. Lagoons of PFA may form a surface crust that appears to be quite stable, and underground fires in colliery spoil tips may result in the formation of similar crusts in spoil deposits.

Pollution of ground water

Introduction

With the exceptions of china clay sand and slate waste, many of the other materials considered in this book may contain traces of toxic compounds that, given the right conditions, could leach out and pollute watercourses. The possibility of this depends on the concentration of the toxic substance, the quantity of material being used and the readiness with which it can be brought into solution.

Materials bound with bitumen or cement are less likely to present any problems for two reasons. First, the particles are encapsulated by a matrix of bitumen or cement which impedes, even if it does not completely prevent, the passage of water into the individual particles. Second, bound materials are mainly used in the upper layers of the road structure in layers that are thin compared with the thicknesses of the underlying layers (see Fig. 4). Apart from the fact that the rate of leaching is likely to be low, the total amount of toxic material that could be leached out is also likely to be low. A similar argument can be applied to unbound layers provided that they are not used in extensive thicknesses and they are not in a permanently wet condition.

However, the lower layers of the road structure may contain considerable volumes of material. This is particularly true of the volume of material in road embankments which cross low-lying areas where high embankments may be required and the material is in a permanently saturated condition and where drainage is poor. Due to the large volumes of embankment material that may be used the amount of toxic material present may become significant even if its concentration is low. This is generally a localized problem because the material is likely to be at least partially encapsulated by materials of very low permeability and the rate of leaching is slow. Moreover, the subsequent leachate is diluted by heavy rainfall so that any toxic effect of the leached compounds is reduced to less than the threshold for toxicity.

Problems could arise, however, where the drainage from the road embankments discharges directly to rivers as pollution of the water may seriously affect aquatic life. Further problems may arise if the rivers receiving discharges are sources of public water supply, or where embankments are constructed close to springs or ground water sources similarly used for the public supply. Although in such circumstances the probability of toxic effects from leached materials would be low, leached materials may give rise to complaints of taste and odour of the water.

Leachates
Leachates can be produced by the action of water on materials containing soluble compounds. Road engineers have always been concerned about the migration of compounds, such as soluble sulphates, that could leach from the unbound pavement layers and subgrade and damage concrete pavements and structures. The Specification for Highway Works contains recommendations for restricting the risk of sulphate attack, which is the main problem as far as the use of naturally-occurring materials is concerned.

However, the chemical composition of alternative materials is much more heterogeneous and they may contain compounds that will leach out and subsequently infiltrate ground water systems and rivers where they may present a danger to plants and crops, fish and other animals. In many instances materials considered to be suitable for re-use have already been used in road or other civil engineering construction works and are known not to produce hazardous leachates. But if materials are produced as a by-product from a chemical, incineration or other process they might produce leachates when exposed to water. Any material suspected of producing hazardous leachates should be investigated before being used in foundation or structural layers of the road. The user should also be satisfied that the material is chemically stable and inert. In case of doubt material suppliers should be asked to provide a chemical analysis of the product to give an assurance that potentially hazardous leachates will not be produced. This information should be confirmed by the waste management division of the local authority. If this information is not available from the supplier the material should not be used until ongoing research investigating leachates has produced more information on which to base decisions or until organizations like the Environment Agency or the DTLR give assurances that the material presents no environmental hazards.

Total chemical analysis of a product is a poor guide to possible pollution problems since it does not indicate the form in which the elements are present. BS EN 1744-3 'Tests for chemical properties of aggregates: Part 3 – Preparation of eluates by leaching of aggregates' describes a method for preparing a sample for analysis. And prEN 12506 'Characterisation of waste – Chemical analysis of eluates' gives the test methods for determining pH and As, Cd, Cr, Va, Cu, Ni, Pb, Zn, Cl, NO_3 and SO_4 contents. Analysis by these methods will give some indication of possible problems but hydrological conditions are important in determining the degree of leaching that may occur in practice. The greatest risk occurs when the unprotected surface is exposed to rainfall during construction or when the road surface is allowed to become badly deteriorated or excessively cracked. The permeability of the material and the surrounding media is important in determining the water regime under which the material will be leached and the leachate dilution.

Research projects are in progress to investigate the potential leachates from construction materials and potential for movement of the leachates through the ground into watercourses. Baldwin *et al.* (1997) examined the leaching potential of nearly all the materials considered in this book. They found that many of the materials tested contained significant amounts of rapidly leached alkali metal ions (sodium and potassium) whereas the leaching of the alkaline earth ions (magnesium, calcium and barium) appeared to be controlled by the limited solubility of these species – their concentration in the leachate might not be high but could continue over a longer period. Analyses were made for a large number of other elements and compounds likely to be present and cadmium and mercury were found to be present in the leachate of several materials.

Baldwin *et al.* (1997) concluded that the potential impacts of using the alternative materials considered by them (which included nearly all the materials covered by this book) were within the range permitted by the strictest water quality standards. They divided their materials into three groups as indicated in Table 42; PFA is not mentioned in the Table because its exceptionally low permeability made it difficult to test.

A collaborative research project sponsored by a part of the European Commission known as ALT-MAT (ALTernative MATerials in road construction) is aimed at providing a toolkit of test methods for mechanical, leaching and hydrodynamic properties. This ongoing work, which has been summarized by Reid (2000a), included an inspection of roads where crushed concrete, demolition rubble, MSWI bottom ash and blast-furnace slag had been used as unbound granular sub-base and capping. Chemical tests revealed that leaching had caused a detectable increase in the concentration of certain constituents of the subgrade below the alternative materials. This was noted in Denmark and France below MSWI bottom ash and in the UK below demolition rubble. However, leaching tests and ground water sampling indicated that the alternative materials did not appear to be having any significant effect on ground water quality.

Table 42. Allocation to application groups for unbound permeable by-products based on leaching tests (Baldwin et al. 1997)

Recommended category	By-product considered
Group 1 (Unrestricted use – similar to limestone)	China clay sand Blast-furnace slag BOS steel slag Colliery spoil Rubber crumb Slate waste Spent oil shale
Group 2 (Some restrictions may apply for unbound material)	Blacktop planings Brick rubble Crushed concrete EAF steel slag
Group 3 (Some restrictions will apply)	MSWI ash

Pollution of the atmosphere

Problems arising from air pollution due to burning colliery spoil have already been considered. The only other type of major air pollution which can arise is from dust, which can be troublesome in windy weather, particularly with small particles of light material such as PFA. It is therefore important that materials should be transported in sheeted lorries and be laid and compacted on arrival on site. Residual problems can be coped with by spraying the compacted material with water and keeping traffic off the surface.

PART 4
REFERENCES

Note: all references to British Standards, European Standards, etc. are included in the text and are not repeated here. The principal English language standards are those published by the British Standards Institution (BSI), the American Society for Testing and Materials (ASTM), the American Association of State Highway and Transportation Officials (AASHTO) and the Comité Européen Normalisation (CEN). Details of these standards can be found in their web pages at:

www.bsi-global.com/standards + commercial
www.astm.org
www.aashto.org
www.cenorm.be

AAS (1999). China clay by-products as aggregates. *Aggregate Advisory Service Digest*, **055**, 1999.

ADAMS J. (1991). *Determined to dig – the role of aggregates demand forecasting in national minerals planning guidance.* Council for the Protection of Rural England, London.

ADVISORY COMMITTEE ON AGGREGATES (1976). *Aggregates: the way ahead.* HMSO, London.

ALT-MAT (1999). *Deliverable D4 Interim Report. Volume 1.* Report No. WP1.SGI.005

ANNUAL ABSTRACT OF STATISTICS (2000). HMSO, London.

BACMI (1991). *The occurrence of utilisation of mineral and construction wastes: A BACMI response to the report by Arup Economics commissioned by DOE.* BACMI, London.

BALDWIN G., ADDIS R., CLARK J. and ROSEVEAR A. (1997). *Use of industrial by-products in road construction – water quality effects.* CIRIA Report 167.

BAMFORTH P.B. (1992). (DHIR R.K. and JONES M.R. (eds)). *What PFA does to concrete.* National seminar on the use of PFA in construction, Dundee, pp. 145–156.

BICZYSKO S. (1999). Reusable materials and recycling for road network rehabilitation. *Quarry Management*, January 1999.

BLEWETT J. and WOODWARD P.K. (2000). Some geotechnical properties of waste glass. *Ground Eng.*, April 2000.

BOND R. (2000). EA probes Newcastle's use of toxic ash in footpaths. *Surveyor*, 18 May 2000.

BRE (1998). Recycled aggregates. *BRE Digest*, 433.

BRITISH GEOLOGICAL SURVEY (2000). *United Kingdom minerals yearbook 1999.* British Geological Survey, Keyworth.

BULLAS J.C. and WEST G. (1991). *Specifying clean, hard and durable aggregate for bitumen macadam roadbase.* Transport and Road Research Laboratory, Crowthorne, UK, Research Report 284.

BURNS J. (1978). *The use of waste and low-grade materials in road construction 6. Spent oil shale.* Transport and Road Research Laboratory, Crowthorne, UK, Report LR 818.

CARR C.E. and WITHERS N.J. (1987). (RAINBOW A.K.M (ed.)) The wetting expansion of cement-stabilized minestone – an investigation of the causes and ways of reducing the problem. *Reclamation, treatment and utilization of coal mining wastes.* Elsevier, Amsterdam, pp. 545–559.

CARSWELL J. and JENKINS E. J. (1996). *Re-use of scrap tyres in highway drainage.* Transport Research Laboratory, Crowthorne, TRL Report 200.

CEGB (1972). *PFA utilization.* Central Electricity Generating Board, London.

CLARKE B. (1992). (DHIR R.K. and JONES M.R. (eds)) *Structural fill.* National seminar on the use of PFA in construction, Dundee, pp. 21–32.

COLLINS R.J. (1990). Case studies of floor heave due to microbiological activity in pyritic shales. *Proc. Conf. on Microbiology in Civil Engineering.* E. & F. N. Spon, London, pp. 288–295.

COLLINS R.J. (1993). Personal communication.

COLLINS R.J. and SHERWOOD P.T. (1995). *Use of wastes and recycled materials as aggregates: standards and specifications.* HMSO, London.

COLLINS R.J., SHERWOOD P.T. and RUSSELL A.D. (1993). *Efficient use of aggregates and bulk construction materials. Volume 1: an overview. Volume 2: technical data and results of surveys.* Building Research Establishment, Garston, UK, Reports BR243 and BR244.

CONCRETE SOCIETY (1998). *Alkali-silica reaction – minimizing the risk of damage to concrete.* Concrete Society, Technical Report No. 30.

CORNELIUS P.D.M. and EDWARDS A.C. (1991). *Assessment of the performance of offsite recycled bituminous material.* Transport and Road Research Laboratory, Crowthorne, UK, Research Report LR 305.

COVENTRY S., WOOLVERIDGE C. and HILLIER S. (1999). *The reclaimed and recycled construction materials handbook.* Construction Industry Research and Information Association, London.

CPRE (1993). *Driven to dig – road building and aggregates demand.* Council for the Protection of Rural England, London.

CPRE (2000). *Minerals planning guidance: aggregate minerals.* Council for the Protection of Rural England, London.

CROCKETT R.N. (1975). *Slate – mineral dossier No. 12.* Mineral Resources Consultative Committee, HMSO, London.

DAWSON A.R. and BULLEN D. (1991). Furnace bottom ash: its engineering properties and its use as a sub-base material. *Proc. Instn. Civ. Engrs.* Part 1, **90**, 993–1009.

DEPARTMENT OF THE ENVIRONMENT (1991). *Environmental effects of surface mineral workings.* HMSO, London.

DEPARTMENT OF THE ENVIRONMENT (1992). *The Environmental Protection Act 1990 – parts II and IV – the controlled waste regulations 1992.* DOE Circular 14/92. HMSO, London.

DEPARTMENT OF THE ENVIRONMENT (1994a). *Sustainable development – The UK strategy. Summary report.* DOE, London.

DEPARTMENT OF THE ENVIRONMENT (1994b). *Guidelines for aggregates provision in England, minerals policy guidance note MPG 6.* HMSO, London.

DESIGN MANUAL FOR ROADS AND BRIDGES (DMRB). Joint publication of the Highways Agency, Scottish Office, Welsh Office and Department of the Environment for Northern Ireland. Volume 1 (1998). *Specification for Highway Works.* Volume 7, Section 1, Part 2 (1995). *Conservation and the use of reclaimed materials in road construction and maintenance.*

DETR (2000). *Waste Strategy for England and Wales.* Department of the Environment, Transport and the Regions/Welsh Office.

DETR (2001a). *Construction and Demolition Waste Survey: England and Wales 1999/2000.* Department of the Environment, Transport and the Regions/Welsh Office.

DETR (2001b). *Waste Strategy – Report of the Market Development Group,* Department of the Environment, Transport and the Regions.

DOE/DOT (1992). *Joint Memorandum by the Departments of the Environment and Transport to the Royal Commission on Environmental Pollution Transport and Environment Study.*

DUNSTAN M.R.H. (1981). *Rolled concrete for dams – a laboratory study of the properties of high fly-ash concrete.* Construction Industry Research and Information Association, London, Technical Note 105.

EARLAND M.G. and MILTON L.J. (1999). *Recycling – structural sustainability.* Conference at TRL, Crowthorne, December 1999.

EMERY J.J. (1976). *Steel slag applications in highway construction.* Paper to the International Conference on the Utilisation of Slags, Mons, Belgium.

ENVIRONMENT AGENCY (1998). *Agency drives home message over waste tyres threat.* Environment Agency Press Release 173/98.

EUROPEAN COMMISSION (2001). *Unbound granular materials for road pavements* Final Report of the COST 337 Action. DG TrEN, Brussels.

FRANKLIN R.E., GIBBS W. and SHERWOOD P.T. (1982). *The use of pulverized fuel ash in lean concrete. Part I – laboratory studies.* Transport and Road Research Laboratory, Crowthorne, UK, Supplementary Report SR 736.

FRASER C.K. and LAKE J.R. (1967). *A laboratory investigation of the physical and chemical properties of burnt colliery shale.* Road Research Laboratory, Crowthome, UK, Report LR 125.

GASPAR L. (1976). Les cendres volants et le laitier granulé en construction routière. *Bulletin de Liaison,* **86,** 135–143.

GOULDEN E.R. (1992). *Slate waste aggregate for unbound sub-base layers.* University of Nottingham, MSc thesis.

GUTT W., NIXON P.J., SMITH M.A., HARRISON W.H. and RUSSELL A.D. (1974). *A survey of the location, disposal and prospective uses of the major industrial by-products and waste materials.* Building Research Establishment, Garston, UK. Current Paper CP/74.

HARDING H.M. and POTTER J.F. (1985). *The use of pulverised fuel ash in lean concrete. Part 2 – pilot-scale trials.* Transport and Road Research Laboratory, Crowthorne, UK, Supplementary Report SR 838.

HAWKINS A.B. and PINCHES G.M. (1987). Cause and significance of heave at Llandough Hospital, Cardiff – a case history of ground floor heave due to gypsum growth. *Quart. J. Eng. Geol.* **20,** 41–57.

HEWLETT P.C. (ed.) (1998). *Lea's chemistry of cement and concrete.* Fourth edition, Edward Arnold, London.

HICKS B. (1999). *In-situ recycling.* Paper given at a Conference on Recycling at the Transport Research Laboratory, Crowthorne, December 1999.

HOCKING R.N. (1994). *China clay wastes.* Seminar on the use and improvement of marginal and waste materials. Geological Society, London.

HOSKING J.R. and TUBEY L.W. (1969). *Research on low-grade and unsound aggregates.* Road Research Laboratory, Crowthorne, UK, Report LR 293.

HOUSE OF COMMONS (2001). *Delivering sustainable waste management.* Report of the Environment, Transport and Regional Affairs Committee. Stationery Office, London 2001

HOWARD HUMPHREYS and PARTNERS (1994). *Managing demolition and construction wastes.* Report of the study on the recycling of construction and demolition wastes in the UK. HMSO, London.

HUBERT P.A. (1987). *Energy use comparison between conventional reconstruction and hot drum mix recycling for major trunk roads.* Proc. seminar highway construction and maintenance. PTRC, London.

KENT COUNTY COUNCIL (1985). *Road trial of phosphoric slag as roadbase.* (Unpublished report prepared for Civil and Marine Ltd).

KETTLE R.J. and WILLIAMS R.I.T. (1978). *Colliery shale as a construction material.* International conference on the use of by-products and waste in civil engineering. Paris.

KWAN J.C.T. and JARDINE F.M. (1997) Ground engineering spoil: practices of disposal and re-use. In *Geoenvironmental Engineering*, Thomas Telford, London.

LEA F.M. (1970). *The chemistry of cement and concrete.* Third Edition, Edward Arnold, London.

LEE A.R. (1974). *Blastfurnace and steel slag.* Edward Arnold, London.

MALLETT S.H., WOOLVERIDGE A.C., TOLLIT H. and BURNETT J.S. (1997). The use of reclaimed aggregate materials in road construction: options for government policy. In *Geoenvironmental Engineering*, Thomas Telford, London.

MEARS (1975). *Mears use slate waste as sub-base on North Wales road contract.* Mears Construction Ltd, Press Release.

MILTON L. J. and EARLAND M.G. (1999). *Design guide and specification for structural maintenance of highway pavements by cold in-situ recycling.* TRL Report 386, 1999

MINISTERE DES TRANSPORTS (1980). *La technique Francaise des assises de chaussées traitees aux liants hydraulique et pozzonaniques.* Ministère des Transports, Direction des Routes et de la Circulation routière, Paris (English translation).

MULHERON M. (1991). *Recycled demolition waste.* Unbound Aggregates in Construction, Nottingham.

NATIONAL COAL BOARD (1983). *Cement bound minestone – user's guide for pavement construction.* NCB Minestone Executive, Whitburn, UK.

NEW CIVIL ENGINEER (1980). Sad Canterbury tale follows minestone's early success. *New Civil Engineer*, 27 November 1980.

NICHOLLS J.C. (2000). *The use of crushed glass in macadam for roadbase layers.* TRL Unpublished Project Report PR/IP/12/00.

NITRR (1986). *Cementitious stabilizers in road construction.* National Institute for Transport and Road Research, CSIR, South Africa, TRH 13.

NIXON P.J. (1978) Floor heave in buildings due to the use of pyritic shales as fill material. *Chemistry and Industry*, 4 March, 1978, pp. 160–164.

OECD (1977). *Use of waste materials and by-products in road construction.* Organisation for Economic Co-Operation and Development, Paris.

OECD (1984). *Road binders and energy savings.* Organisation for Economic Co-Operation and Development, Paris.

OECD (1997). *Recycling strategies for road works.* Organisation for Economic Co-Operation and Development, Paris.

O'MAHONEY M.M. (1990). Recycling of materials in civil engineering. University of Oxford, PhD thesis.

PLEASE A. and PIKE D.C. (1968). *The demand for road aggregates.* Transport and Road Research Laboratory, Crowthorne, UK, Report LR 185.

PORTLAND CEMENT ASSOCIATION (1979). *Soil-cement construction handbook.* Portland Cement Association, Skokie, Illinois.

POTTER J.F. (1996). *Road haunches: a guide to reusable materials.* Transport Research Laboratory, Crowthorne, UK, Report No. 216.

POTTER J.F., SHERWOOD P.T. and O'CONNER M.G.D. (1985). *The use of pulverised fuel ash in lean concrete. Part 3 – field studies.* Transport and Road Research Laboratory, Crowthorne, UK, Supplementary Report SR 842.

QPA (2001). *Aggregates tax is a £385 million threat to construction.* Quarry Products Association Press Release, March 2001.

RAINBOW A.K.M. (1989). *Geotechnical properties of United Kingdom minestone.* British Coal, London.

REID J.M. (2000a). *The use of alternative materials in road construction.* Paper presented to UNBAR Conference, Nottingham University.

REID J.M. (2000b). Disarming the resistance. *Ground Engineering,* December 2000.

RILEM (1994) Specifications for concrete with recycled aggregates. *Materials and Structures,* **27**, 557–559.

ROCKLIFF D. (1998). Appropriate use of low-grade aggregates in highway works. *Quarry Management,* May 1998.

ROE P.G. (1976). *The use of waste and low-grade materials in road construction 4. Incinerated refuse.* Transport and Road Research Laboratory, Crowthorne, UK, Report LR 728.

ROYAL COMMISSION ON ENVIRONMENTAL POLLUTION (1974). *Fourth report – Pollution control: progress and problems.* HMSO, London.

ROYAL COMMISSION ON ENVIRONMENTAL POLLUTION (1993). *Seventeenth report – Incineration of waste.* HMSO, London.

ROYAL COMMISSION ON ENVIRONMENTAL POLLUTION (1994). *Eighteenth report – Transport and the environment.* HMSO, London.

SHERWOOD P.T. (1987). Wastes for imported fill. *ICE Construction Works Guide.* Thomas Telford, London.

SHERWOOD P.T. (1993). *Soil stabilization with cement and lime – a state of the art review.* HMSO, London.

SHERWOOD P.T. and ROE P.G. (1974). *The effect on the landscape of borrow-pits used in major roadworks.* Transport and Road Research Laboratory, Crowthorne, UK, Supplementary Report 122 UC.

SHERWOOD P.T. and RYLEY M.D. (1966). *The use of stabilized pulverized fuel ash in road construction.* Road Research Laboratory, Crowthorne, UK, Report LR 49.

SHERWOOD P.T. and RYLEY M.D. (1970). *The effect of sulphates in colliery shale on its use for roadmaking.* Road Research Laboratory, Crowthorne, UK, Report LR 324.

SWEERE G.T.H. (1989) (JONES R.H. and DAWSON A.R. (eds)) Structural contribution of self-cementing granular bases to asphalt pavements. *Unbound aggregates in roads.* Butterworths, London, pp. 343–353.

SWEERE G.T.H. (1991). *Re-use of demolition waste in road construction.* Unbound Aggregates in Construction, Nottingham University, Nottingham.

TANFIELD D.A. (1978). *The use of cement-stabilized colliery spoils in pavement construction.* International conference on the use of by-products and waste in civil engineering, Paris.

THOMAS M.D.A., KETTLE R.J. and MORTON J.A. (1987). (RAINBOW A.K.M. (ed.)) Short-term durability of cement-stabilized minestone. *Reclamation, treatment and utilization of coal mining wastes.* Elsevier, Amsterdam, pp. 533–544.

THOMPSON P.D. and SZATKOWSKI W. S. (1971) *Full-scale road experiments using rubberized surfacing materials.* Road Research Laboratory, Crowthorne, UK, Report LR370.

TUBEY L.W. (1978). *The use of waste and low-grade materials in road construction – 5. China clay sand.* Transport and Road Research Laboratory, Crowthorne, UK, Report LR 819.

TUBEY L.W. and WEBSTER D.C. (1978). *Effects of mica on the roadmaking properties of materials.* Transport and Road Research Laboratory, Crowthorne, UK, Supplementary Report 408.

UKQAA (1998a). *Fly ash bound mixtures (FABM) for road and airfield pavements.* UKQAA Technical Data Sheet 6.0.

UKQAA (1998b). PFA, FBA and the control of substances hazardous to health regulations 1994. UKQAA Health and Safety Information Sheet H.

VERHASSELT A. and CHOQUET F. (1989). (JONES R.H. and DAWSON A.R. (eds)) Steel slags as unbound aggregate in road construction: problems and recommendations. *Unbound aggregates in roads.* Butterworths, London, pp. 204–209.

WATSON K.L. (1980). *Slate waste – engineering and environmental aspects.* Applied Science Publishers, London.

WEST G. and O'REILLY M.P. (1986). *An evaluation of unburnt colliery shale as fill for reinforced earth structures.* Transport and Road Research Laboratory, Crowthorne, UK, Research Report 97.

WHITBREAD M., MARSEY A. and TUNNELL C. (1991). *Occurrence and the utilisation of mineral and construction wastes.* Report for the Department of the Environment. HMSO, London

WOOD C.E.J. (1998). A specification for lime-stabilization of subgrades. *Lime stabilization '88'.* BACMI, London, pp. 2–8.

Index